WISCONSIN, ILLINOIS & IOWA

Rocks & Minerals

A Field Guide to the Badger, Prairie & Hawkeye States

Dan R. Lynch & Bob Lynch

Adventure Publications, Inc.
Camb[illegible]

Dedication

To Nancy Lynch, wife of Bob and mother of Dan, for her love and continued support of our book projects.

And to Julie, Dan's wife, for her love and patience.

Acknowledgments

Thanks to the following for providing specimens and/or information: Phil Burgess, Kevin Ponzio, Rob Carlson, Nevin Franke and Jeff Theroux

Photography by Dan R. Lynch

Cover and book design by Jonathan Norberg

Edited by Brett Ortler

10 9 8 7 6 5 4 3

Rocks & Minerals of Wisconsin, Illinois & Iowa

Published by Adventure Publications
An imprint of AdventureKEEN
310 Garfield Street South
Cambridge, Minnesota 55008
(800) 678-7006
www.adventurepublications.net

Printed in China
ISBN 978-1-59193-451-6 (pbk.)

Table of Contents

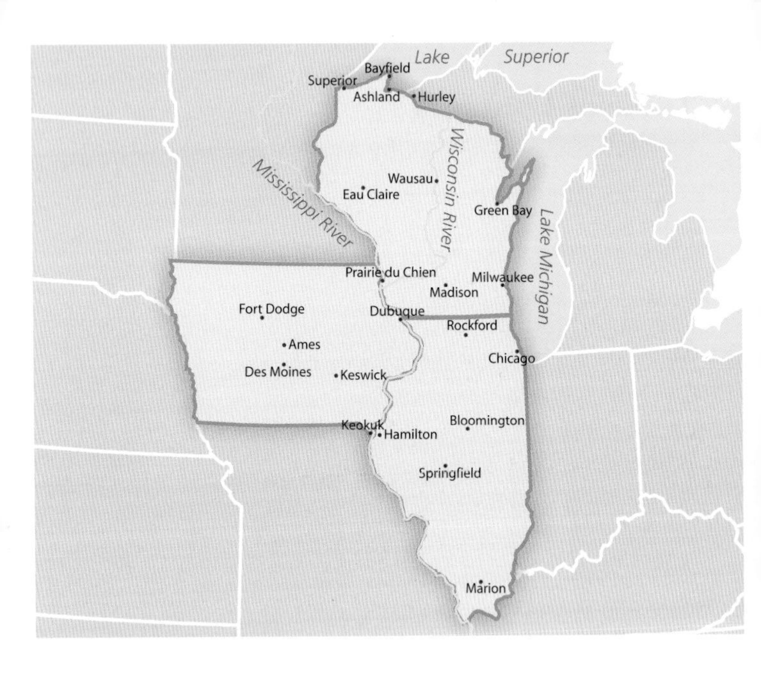

Introduction

Wisconsin, Iowa and Illinois make up much of the Upper Midwest and boast some of the best rock and mineral collecting in the region. These states' beautiful river valleys and rolling hills are home to scientifically important plant and animal fossils, world-class mineral specimens, and a unique assemblage of rocks of all kinds. With a rich and ongoing history of mining, each state continues to produce new finds for collectors. This book will discuss many of the rocks and minerals that these three states have to offer, including everything from their most common rocks to their rarest crystals. More importantly, we'll tell you what to look for and how to identify what you've found.

Important Terms and Definitions

Books about geology can be hard for some hobbyists to understand. So in order to make this book intuitive for novices yet still useful for experienced collectors, we have included a few technical geological terms in the text, but we "translate" them immediately after using them by providing a brief definition. In this way, amateurs can learn some of the more important terms relevant to the hobby in an easy, straightforward manner. Of course, all of the geology-related terms used here are defined in the glossary found at the back of this book as well. But for those entirely new to rock and mineral collecting, there are a few very important terms you should understand not only before you begin researching and collecting minerals, but even before you read this book.

Many people go hunting for rocks and minerals without knowing the difference between the two. But the difference is simple. A mineral is formed from the crystallization (solidification) of a chemical compound, which is a combination of elements. For example, silicon dioxide, a chemical compound consisting of the elements silicon and oxygen, crystallizes to form quartz, the most abundant mineral on earth. In contrast, a rock is a mass of solid material containing a mixture of many different minerals. While pure minerals exhibit very definite and testable characteristics, such as a distinct, repeatable shape and hardness, rocks do not and vary greatly because they consist of various minerals. For amateurs, this can make rocks harder to identify than minerals.

Many of the important rockhounding terms apply only to minerals and their crystals. A crystal is a solid object with a distinct shape and a repeating atomic structure created when a chemical compound solidified. In other words, when different elements come together, they form a chemical compound that takes on a specific shape when it hardens. For example, the mineral galena is lead sulfide, a chemical compound consisting of lead and sulfur, which crystallizes, or solidifies, into the shape of a cube. A "repeating atomic structure" means that when a crystal grows, it builds upon itself. If you compared two crystals of galena, one an inch long and the other a foot long, they would have the same identical cubic shape. In contrast, if a mineral is not found in a well-crystallized form but rather as a solid, rough chunk, it is said to be massive. If a mineral typically forms massively, it will frequently be found as irregular pieces or masses, rather than as well-formed crystals.

Cleavage is the property of some minerals to break in a particular way when carefully struck. As solid as minerals may seem, many have planes of weakness within them that derive from a mineral's internal crystal structure. These points of weakness are called cleavage planes and it is along these planes that some minerals will cleave, or separate, when struck. For example, galena has cubic cleavage, and even the most irregular piece of galena will fragment into perfect cubes if carefully broken.

Luster is the intensity with which a mineral reflects light. The luster of a mineral is described by comparing its reflectivity to that of a known material. A mineral with "glassy" luster (also called a "vitreous" luster) is as "shiny" as glass. The most poorly reflective minerals have a "dull" luster, while "adamantine" describes the most brilliant. Minerals with a "metallic" luster clearly resemble metal, and this can be a very diagnostic trait. But determining a mineral's luster is a subjective experience, so not all observers will necessarily agree on luster, especially when it comes to less obvious lusters, such as "waxy," "greasy" or "earthy."

When minerals form, they do so on, or in, rocks. Therefore, if you hope to successfully find a specific mineral you need to be able to distinguish between the different types of rocks that exist. Igneous rocks form as a result of volcanic activity and originate from magma, lava or volcanic ash. Magma is hot, molten rock buried deep within the earth; it can take extremely long periods of time for it to cool and form rock. Lava, on the other hand, is molten rock that has reached the earth's surface, where it cools and solidifies into rock very rapidly. Sedimentary rocks typically form at the bottoms of lakes and oceans when sediment compacts and solidifies into layered masses. This sediment can contain organic matter as well as weathered fragments or grains from broken-down igneous, metamorphic or sedimentary rocks. Finally, metamorphic rocks develop when igneous, sedimentary or even other metamorphic rocks are subjected to heat and/or pressure within the earth. This changes both their appearance and their mineral composition.

A Brief Overview of the Region's Geology

At first glance, the rolling hills of Wisconsin, the farmlands of Iowa, and the plains of Illinois may not seem geologically interesting. But in reality, the rocks that underlie these three states owe their formation to fascinating rock-building events that were crucial to the formation of not only the Upper Midwest, but much of North America as we know it. In fact, if you take a closer look at these states' seemingly mundane landscapes, you can deduce the presence of past mountain ranges, violent volcanic events, and the existence of the many warm, shallow seas that once inundated the region. All of this geological history makes the Upper Midwest a compelling area to study.

Wisconsin's Geological Past

The geological history of a state or region is defined by its bedrock, which is the solid rock found just beneath the loose soil and gravel that sits on the surface. When bedrock forms, it does so in layers, with the youngest layers on the top and the oldest layers deeply buried below. Each layer may be a very different type of rock, and when weathering erodes and wears away the upper, younger layers of bedrock, older layers are exposed, giving us a view of the deeper rock formations. Therefore, depending on erosion, a state's bedrock will change from place to place, and by studying which layers are at the surface and what type of rock they consist of, geologists can begin to piece together the origin of the state's landforms and its geological history.

The history of Wisconsin's bedrock begins nearly three billion years ago during a geological period known as the Archean Eon. At this time, there was no life on land and little more than simple bacteria in the seas, and volcanic activity was much more common than it is today. It was during this turbulent time that a new continent was developing. Known today as the Superior Craton, it consisted largely of granite and existed for hundreds of millions of years before another landmass called the Marshfield Terrane collided with it around 1.9 billion years ago. As these two landmasses pushed into each other, a string of granitic mountains known as the Penokee Range appeared across what is now northern Wisconsin, extending into Minnesota and Michigan. Through the next few hundred million years, volcanic eruptions added more rock to the ancient Wisconsin landscape. About 1.1 billion years ago, the Lake Superior region underwent a catastrophic event known as the Midcontinent Rift. The result of tectonic plates drifting apart, the Midcontinent Rift resulted in a 1,200-mile-long tear in the earth. If the rift had succeeded, it would have formed a sea. Instead, it failed and filled in with lava that cooled and formed thick masses of dense rock. This short summary just scratches the surface of this nearly 2-billion-year-long chapter in Wisconsin's history, but it provides a contrast to the less dramatic but equally interesting eons to come.

The next most notable change to Wisconsin's early landscape occurred around 600 million years ago, when sea levels rose and a warm, shallow sea inundated the entire region. Thick beds of sandstone formed on the seafloor, covering the weathered remains of the Penokee Range as well as the rest of the state. The sea eventually receded, but 490 million years ago another body of water once again submerged Wisconsin. Even shallower and warmer than the previous sea, it was also teeming with life. Limestone, a sedimentary rock that consists of the remains of ancient sea life, was the primary type of rock forming at this time. More seas dominated the region from 430 to 415 million years ago, depositing yet more limestone and forming the majority of the bedrock of modern-day Wisconsin. Rock layers for the next 400 million years are missing, and this indicates that a long period of erosion occurred. Erosion breaks down exposed bedrock and, if uninterrupted by other rock-forming events, can completely "erase" layers of rock, leaving behind only sand.

While dramatic geological events created Wisconsin's rocks and collectible minerals, erosion gave the state many of its modern-day features. The most prominent erosive features were the glaciers of the past ice age. Present from around 2.5 million years ago until just 12,000 years ago, these incredible sheets of ice were thousands of feet thick and utterly transformed the state. They scraped, plowed and sculpted the land, sheared off layers of bedrock, smoothed hills and left behind incredible amounts of gravel, sand and mud (called glacial till). As the glaciers melted, enormous lakes and raging rivers further shaped the land. Today, we see evidence of the glaciers all over Wisconsin in the form of flat plains, lakes, moraines (large dome-like ridges of gravel) and many other features. The exception is the southwestern corner of the state, which is known as the "driftless area" because glacial till was once called glacial drift. This portion of the state never saw glaciation, and it therefore retains a more rugged landscape and is filled with deep valleys and hills.

Because Wisconsin endured hundreds of millions of years of weathering in addition to the glaciers, exposures of 1.1-billion-year-old volcanic rocks can be found in northern Wisconsin. These dark, agate-filled rocks are the "roots" of the Penokee Range in north-central Wisconsin. What's more, erosion has exposed thousands of square miles of ancient seafloor in the central and southern parts of the state. All of this makes Wisconsin a perfect destination for rock hounds.

Iowa's Geological Past

Nearly all of Iowa's bedrock formed between 505 and 74 million years ago—nowhere near as long ago as many of Wisconsin's rocks. They are a result of the same rising and falling seas that repeatedly submerged Wisconsin so long ago. Each subsequent body of water left behind sedimentary rocks, particularly limestone, sandstone and shale, forming thick, fossil-rich layers that are still visible anywhere in Iowa today. No rock-forming events have happened since the last sea receded, but below Iowa's surface lies evidence of a more dramatic geological past: magnetic imaging shows that deep below the sedimentary bedrock of Iowa lies a huge portion of the Midcontinent Rift that tore open the continent over a billion years ago.

When the glaciers of the last ice age descended from Canada, they repeatedly reached Iowa, scraping and smoothing the land and covering the entire state with a thick layer of glacial till, or gravel. But while the glaciers advanced and retreated many, many times, not all parts of Iowa saw the same amount of glaciation. For example, nearly the entire southern half of the state today is characterized by gentle rolling hills and river valleys, while the north is comparatively flat and smooth.

The last glacier that reached southern Iowa retreated around 500,000 years ago, leaving the vast expanses of glacial till to erode and weather as rain, wind and rivers cut through them. In the north, however, the most recent glacier retreated just 12,000 years ago, so the remaining till there has had far less time to be sculpted by the elements. In northeastern and southeastern Iowa, the Mississippi River has weathered away much of the till, revealing bedrock, as has the Missouri River in the southwestern portion of the state. Most everywhere else, copious amounts of glacial gravel remain in various states of erosion.

Iowa's bedrock also holds evidence of some of the most violent and sudden events in earth's history: meteor impacts. Several impact structures are present in the state, but the most significant is the Manson Impact Structure in north-central Iowa. Believed to have been made by an asteroid 74 million years ago, the 24-mile-wide crater is filled with rock and completely buried under glacial till today, but the presence of breccia (a rock consisting of broken stone fragments) in the crater is an abundant indication of the impact's incredible power.

Illinois' Geological Past

As in Iowa, inland seas played a large role in the history of Illinois, forming the basis of the state's sedimentary bedrock. But in Illinois, seawater was concentrated in a specific geological structure: the Illinois Basin. The basin, surrounded by high-elevation features, such as Missouri's Ozark Mountains, is the dominant feature of the state and extends into Kentucky and Indiana. Around 400 million years ago, enormous amounts of limestone, sandstone and shale were beginning to form in the basin, and large amounts of organic material was buried along with these rocks. Over the eons, the dead plants and animals turned into coal, hydrocarbons and oil, making the Illinois Basin a major fuel production area today. In addition, the coal and oil naturally released sulfur into the surrounding limestone.

The limestone itself began as ocean sediment consisting largely of the shells and other hard parts of aquatic organisms. Initially, these hard remains were composed of aragonite, a calcium-bearing mineral produced by many aquatic lifeforms, but due to pressure and chemical forces, the aragonite sediments recrystallized into calcite. This process solidified all the sediment together to form limestone, but it also released zinc and lead that were originally trapped in the aragonite as impurities. These metals were then free to bond with the sulfur released from the area's coal and oil, creating new minerals in concentrations large enough to make southern Illinois a significant lead and zinc mining region.

While most of Illinois is underlain by largely flat sedimentary rock, this bedrock is not visible in most parts of the state due to the incredible amounts of glacial till present, especially in the north-central part of the state where deposits of gravel are over 500 feet deep. As you travel southward, however, there is less and less till, until you reach the southern tip of Illinois, an area where glaciers never reached. Here, some of the youngest bedrock material in the state is still intact and exposed.

Precautions and Preparations

It may be surprising, but rock and mineral collecting often brings with it several dangers and legal concerns. As a collector, it is always your responsibility to know where you can legally collect, which minerals may be hazardous to your health, and what you need to take with you in order to be prepared for any difficulties you may face. Here we will detail some of these issues.

Protected and Private Land

Wisconsin, Iowa and Illinois have nationally protected parks and monuments as well as American Indian reservations, all of which are areas where it is illegal to collect anything. In the region's many state parks, collecting is generally prohibited; in Wisconsin, no collecting is allowed in any state park or forest, and in Iowa or Illinois, certain parks may allow collecting but only for those with the proper permits. We encourage collectors to obey the law and leave the designated natural spaces wild and untouched for generations to come.

As in any state, many places in Wisconsin, Iowa and Illinois are privately owned, including wilderness areas that may not have obvious signage. Needless to say, you are trespassing if you collect on private property and the penalty may be worse than just a fine. In addition, property lines (and owners) change frequently, so just because a landowner gave you permission to collect on their property last year doesn't mean the new owner will like you on their property this year. Always be aware of where you are.

Quarries, Rivers, Sinkholes and Snakes

When in the field, collectors need to remain vigilant and cautious to remain safe. Many quarries, gravel pits and mine dumps present amazing collecting opportunities, but they may have large rock piles or pit walls that are unstable and prone to collapse. Never go beneath overhanging rock, and keep clear of unstable rock walls. Rivers also present their own dangers, and even though the water's surface may look calm, strong currents may be present. It doesn't take very much current to make you lose your footing completely. In limestone-rich areas such as Wisconsin, Iowa and Illinois, sinkholes can occur, and small ones may be hidden by grass and other plants, presenting a tripping or falling hazard. Finally, bear in mind the possibility of encountering dangerous wildlife, particularly snakes. The region is home to a few species of venomous rattlesnakes, which often hide under rocks and in holes a rock hound may wish to inspect. Always be cautious when moving large stones and never reach blindly into dark spaces, and certainly never try to move or otherwise provoke a snake if you find one. In all cases, vigilance and care will ensure a safe collecting trip.

Equipment and Supplies

When you set out to collect rocks and minerals, there are a few items you don't want to forget. No matter where you are collecting, leather gloves are a good idea, as are knee pads if you plan to spend a lot of time on the ground. If you think you'll be breaking rock, bring your rock hammer (not a nail hammer) and eye protection. If the weather is hot and sunny, take the proper precautions and use sunblock, wear sunglasses and a hat, and bring ample water, both for drinking and for rinsing specimens. Lastly, bringing a global positioning system (GPS) device is a great way to prevent getting lost.

Collecting Etiquette

Too often, popular collecting sites are closed by their landowners due to litter, trespassing and vandalism. In many of these cases, the landowners may have been kind enough to allow collectors onto their land, but when people trespass rather than simply ask for permission, we all lose. When collecting, never go onto private property unless you've obtained permission, and be courteous; don't dig indiscriminately or take more than you need. If you follow these rules, you're likely to be invited back. To ensure great collecting sites for future rock hounds, dig carefully, take only a few specimens, and leave the location cleaner than you found it.

Dangerous Minerals and Protected Fossils

Potentially Hazardous Minerals

The vast majority of minerals in the three-state region are completely safe to handle, collect and store, but a few do have some inherent dangers associated with them. Potentially hazardous minerals included in this book are identified with the symbol shown above. Always take proper precautions with these minerals.

Amphibole group (page 45)—a few varieties are asbestos; asbestiform minerals form as delicate, flexible fiber-like crystals that can easily become airborne and inhaled, posing a cancer risk; these varieties are very rare in this region

Cerussite (page 71)—contains lead, a toxin; wash hands after handling

Galena (page 127)—contains lead, a toxin; wash hands after handling

It should be noted that collectors don't need to shy away from these minerals; you just need to be mindful of their potential hazards. Galena, for example, contains lead, but the lead atoms are bonded to sulfur atoms, making the mineral safe to handle—but you probably will still want to wash your hands afterwards. With these minerals (and most other minerals in general), the real danger is in inhaling their dust or ingesting them, which is easy to avoid.

Protected Fossils and Artifacts

Fossils are the remains of ancient plants and animals that have turned to rock. Fossils are extremely popular collectibles, but in most states there are strict rules about what you can and cannot collect. Wisconsin, Iowa and Illinois are no exception.

Vertebrate fossils, or fossils of animals with a backbone, such as fish, are protected due to their possible scientific significance. These should be reported to the Bureau of Land Management, and any collecting or tampering with the fossils may lead to fines or other penalties. In general, if you want to collect on state-owned land, you need a permit. This is especially true for fossils. For more information, contact the Bureau of Land Management's Northeastern States Field Office, which has jurisdiction over Wisconsin, Iowa and Illinois, by calling (414) 297-4400, or by visiting www.blm.gov.

Artifacts created by American Indians may be discovered in Wisconsin, Iowa and Illinois, and this includes arrowheads, pottery, and pottery fragments. It is illegal in all cases to disturb or collect these artifacts! They may hold considerable scientific and cultural value. If you come across an artifact, contact the Bureau of Land Management.

Hardness and Streak

There are two important techniques everyone wishing to identify minerals should know: hardness and streak tests. All minerals will yield results in both tests, as will certain rocks, which makes these tests indispensable to collectors.

The measure of how resistant a mineral is to abrasion is called hardness. The most common hardness scale, called the Mohs Hardness Scale, ranges from 1 to 10, with 10 being the hardest. An example of a mineral with a hardness of 1 is talc; it is a chalky mineral that can easily be scratched by your fingernail. An example of a mineral with a hardness of 10 is diamond, which is the hardest naturally occurring substance on earth and will scratch every other mineral. Most minerals fall somewhere between 2 and 7 on the Mohs Hardness Scale, so learning how to perform a hardness test (also known as a scratch test) is critical. Common tools used in a hardness test include your fingernail, a U.S. nickel (coin), a piece of glass and a steel pocket knife. There are also hardness kits you can purchase that have a tool of each hardness.

To perform a scratch test, you simply scratch a mineral with a tool of a known hardness—for example, we know a steel knife has a hardness of about 5.5. If the mineral is not scratched, you will then move to a tool of greater hardness until the mineral is scratched. If a tool that is 6.5 in hardness scratches your specimen, but a 5.5 did not, you can conclude that your mineral is a 6 in hardness. Two tips to consider: As you will be putting a scratch on the specimen, perform the test on the backside of the piece (or, better yet, on a lower-quality specimen of the same mineral), and start with tools softer in hardness and work your way up. On page 20, you'll find a chart that shows which tools will scratch a mineral of a particular hardness.

The second test every amateur geologist and rock collector should know is streak. When a mineral is crushed or powdered, it will have a distinct color—this color is the same as the streak color. When a mineral is rubbed along a streak plate, it will leave behind a powdery stripe of color, called the streak. This is an important test to perform because sometimes the streak color differs greatly from the mineral itself. Hematite, for example, is a dark, metallic and gray mineral, yet its streak is a rusty red color. Streak plates are sold in some rock and mineral shops, but if you cannot find one, a simple unglazed piece of porcelain from a hardware store will work. But there are two things you need to remember about streak tests: If the mineral is harder than the streak plate, it will not produce a streak and will instead scratch the plate itself. Secondly, don't bother testing rocks for streak; they are made up of many different minerals and won't produce a consistent color.

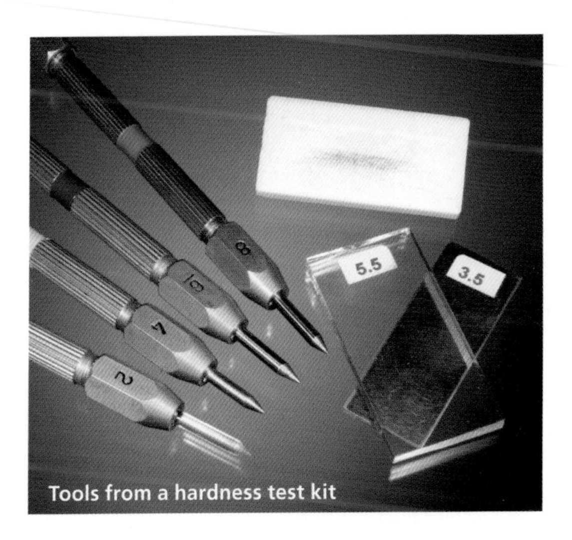

Tools from a hardness test kit

The Mohs Hardness Scale

The Mohs Hardness Scale is the primary measure of mineral hardness. This scale ranges from 1 to 10, from softest to hardest. Ten minerals commonly associated with the scale are listed here, as well as some common tools used to determine a mineral's hardness. If a mineral is scratched by a tool of a known hardness, then you know it is softer than that tool.

HARDNESS	EXAMPLE MINERAL	TOOL
1	Talc	
2	Gypsum	
2.5		Fingernail
3	Calcite	
3.5		U.S. nickel, brass
4	Fluorite	
5	Apatite	
5.5		Glass, steel knife
6	Orthoclase feldspar	
6.5		Streak plate
7	Quartz	
7.5		Hardened steel file
8	Topaz	
9	Corundum	
9.5		Silicon carbide
10	Diamond	

For example, if a mineral is scratched by a U.S. nickel (coin) but not your fingernail, you can conclude that its hardness is 3, equal to that of calcite. If a mineral is harder than 6.5, or the hardness of a streak plate, it will instead scratch the streak plate itself, unless the mineral has weathered to a softer state.

Quick Identification Guide

Use this quick identification guide to help you determine which rock or mineral you may have found. Listed here are the primary color groups followed by some basic characteristics of the rocks and minerals of Wisconsin, Iowa and Illinois, as well as the page number where you can read more. While the most common traits for each rock or mineral are listed here, be aware that your specimen may differ greatly.

WHITE OR COLORLESS

If white or colorless and...	then try...
Soft, light-colored masses that occur with gypsum but are harder	anhydrite, page 47
Glassy, steeply tapered needle-like crystals, often arranged in radial groupings	aragonite, page 49
Tabular (thin, plate-like) glassy crystals, often arranged in parallel groups	barite, page 53
Opaque, intergrown rhombohedral crystals resembling calcite but harder; only found in southern Illinois	Benstonite, page 59
Common, soft, six-sided, tooth-like crystals or blocky masses embedded in rocks like limestone	calcite, page 65
Short, blocky, glassy crystals with complex tips, often found with fluorite in limestone	celestine, page 69
Tiny crystals, bubbly crusts, or powdery masses on the surface of galena	cerussite, page 71

Quick Identification Guide *(continued)*

WHITE OR COLORLESS

(continued)	**If white or colorless and...**	**then try...**
	Soft, often chalky, crusts and coatings found inside geodes	clay, page 85
	Hard, glassy, reflective, blocky masses found in rocks, particularly in granite	feldspar group, page 101
	Glassy, extremely soft crystals easily scratched with your fingernail; found in clay beds or within geodes	gypsum, page 145
	Hard, blocky masses that exhibit an internal "schiller" (flashes of color when rotated in light)	"moonstone," page 175
	Very small, very hard, brightly lustrous crystals found in granite from central Wisconsin	phenakite, page 183
	Hard, abundant, six-sided pointed crystals or light-colored masses embedded in rocks	quartz, page 195
	Very rare clusters of needle-like crystals that are harder than calcite; typically found in southern Illinois	strontianite, page 213
	Radial clusters of soft, needle-like crystals found within cavities in basalt in northern Wisconsin	zeolite group, page 221

Quick Identification Guide (continued)

GRAY

If gray and...	then try...
Very hard, common, opaque masses of rock with waxy surfaces (when weathered); breaks with sharp edges	chert, page 79
Dense, dark rocks with mottled coloration; contains dark glassy crystals; found primarily in northern Wisconsin	diabase or gabbro, page 95
Brittle, slender branching structures found in sand; often tube-like in shape with glassy interiors	fulgurite, page 125
Soft, metallic mineral found in limestone; breaks into perfect cubes when struck	galena, page 127
Shiny, soft, nearly metallic, flexible crystals or masses embedded in coarse-grained rocks or schist	mica group, page 169
Extremely fine-grained rocks that are dense, gritty and somewhat soft and often found with chert	mudstone or siltstone, page 177
Hard grayish-green crystals found in clusters within basalt and other rocks in northern Wisconsin	prehnite, page 187
Soft rocks composed of compressed sand; individual grains of sand are easily freed	sandstone, page 203

Quick Identification Guide (continued)

BLACK

If black and...	then try...
Hard grains or masses, often lustrous with a silky sheen; found embedded in coarse-grained rocks	amphibole group, page 45
Dark, fine-grained rock common in northern Wisconsin, particularly around Lake Superior	basalt, page 55
Soft, dull, cracked crusts or sticky coatings on minerals in limestone, particularly in geodes	bitumen, page 61
Soft, very dark, lightweight and lustrous rocky masses	coal, page 87
Dark, sooty and sometimes metallic crusts or tree-shaped growths on rock surfaces	manganese oxides, page 163
Hard, glassy, dark masses or blocky or needle-like crystals embedded in dark rocks, particularly gabbro	pyroxene group, page 193
Dark masses or veins of very hard, slender, needle-like crystals in granite	schorl (tourmaline), page 217
Fairly hard, dark rock with tight layering	slate, page 205
Soft, dark, complex, generally triangular crystals, often translucent in bright light and found with chalcopyrite	sphalerite, page 211

Quick Identification Guide (continued)

If green or blue and...	then try...
Soft, vividly colored blue crusts or veins associated with copper, malachite or chalcopyrite	azurite, page 51
Very soft dark-green masses; often found as crusts in basalt vesicles or as fine grains in schists	chlorite group, page 81
Soft, chalky or crumbly masses or crusts, usually alongside copper or malachite	chrysocolla, page 83
Hard, yellowish-green crystals, crusts or masses that are glassy or grooved	epidote, page 99
Tiny green grains or stains in sandstone	glauconite, page 133
Soft, vividly colored fibrous crusts or masses; typically associated with copper or chalcopyrite	malachite, page 161
Hard, glassy, translucent grains or masses; typically found within rocks like gabbro, or loose as tiny grains in sand	olivine group, page 179
Hard grayish-green crystals found in clusters within basalt and other rocks in northern Wisconsin	prehnite, page 187
Hard, glassy, dark masses or needle-like crystals embedded in rocks such as pegmatite	pyroxene group, page 193

Quick Identification Guide (continued)

GREEN OR BLUE

(continued) If green or blue and...	then try...
Extremely soft, lustrous, flaky masses that are easily scratched with a fingernail	talc, page 215

TAN TO BROWN

If tan to brown and...	then try...
Chalky crusts of soft, brownish material on the surface of galena	cerussite, page 71
Very hard, translucent, waxy ball-like masses, often with uneven coloration	chalcedony, page 73
Very hard, common, opaque masses of rock with waxy surfaces when weathered; breaks with sharp edges	chert, page 79
Soft, gritty, crumbly material that becomes sticky when wet; typically found along rivers	clay, page 85
Uniquely round, hard, ball- or blob-like masses found in or near sedimentary rocks	concretions, page 89
Soft, blocky, rhombohedral crystals with curved faces and a pearly luster; often found in limestone cavities	dolomite, page 97
Soft, cubic crystals; often occurring with calcite in limestone	fluorite, page 103

Quick Identification Guide *(continued)*

TAN TO BROWN

(continued)	If tan to brown and...	then try...
	Chert or limestone containing coral and other aquatic life	aquatic fossils, page 107
	Chert or limestone containing smooth, domed clam-like shells	bivalve fossils, page 109
	Limestone containing deeply grooved, fan-shaped clam-like shells	brachiopod fossils, page 111
	Limestone containing long, tapering segmented shells, sometimes large in size	cephalopod fossils, page 113
	Chert or limestone containing snail-like shells	gastropod fossils, page 115
	Limestone, shale or mudstone containing plants or plant material	plant fossils, page 117
	Rocks containing banded structures that are wave- or mushroom-like	stromatolite fossils, page 119
	Limestone containing unusual networks of tunnel-like channels	trace fossils, page 121
	Limestone containing fragments of washboard-like segmented shells	trilobite fossils, page 123

Quick Identification Guide (continued)

(continued) If tan to brown and...	then try...
Extremely soft masses that are easy to scratch with your fingernails; sometimes found with a fibrous structure	gypsum, page 145
Soft, widely abundant rock that can easily be scratched with a knife and fizzes in vinegar	limestone, page 155
Common coatings, crusts or veins of rusty or chalky material with no crystal shape	limonite, page 157
Extremely fine-grained rocks that are dense, gritty and somewhat soft; often found with chert	mudstone or siltstone, page 177
Very hard, glassy rock that resembles quartz but has a grainy texture and a flake-like appearance when broken	quartzite, page 199
Rough, often crumbly rock composed of tiny grains of sand cemented together	sandstone, page 203
Soft, fine-grained rock consisting of flat, parallel layers that can often be separated with a knife	shale, page 205
Soft, light-colored blocky or blade-like crystals within cavities in iron-rich rocks or geodes	siderite, page 207
Hard, fibrous clusters of silky hair-like crystals in metamorphic rocks such as schist	sillimanite, page 209

Quick Identification Guide *(continued)*

TAN TO BROWN

(continued)	If tan to brown and...	then try...
	Very hard, tiny, lustrous crystals with triangular faces; found embedded in granite	zircon, page 223

RED TO ORANGE

	If red to orange and...	then try...
	Very hard, translucent, waxy ball-like masses, often with an uneven coloration	chalcedony, page 73
	Hard, glassy, reflective, blocky masses or intergrown crystals found in rocks, particularly in granite and pegmatite	feldspar group, page 101
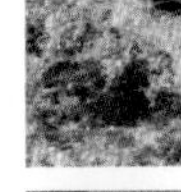	Very hard, rounded, ball-like crystals embedded in granite or schist	garnet group, page 129
	Very hard, opaque masses with a rough texture when broken but a smooth, waxy look and feel when weathered	jasper, page 151
	Light-colored, hard, fine-grained rock often containing faint bands; mostly found in northern Wisconsin	rhyolite, page 201
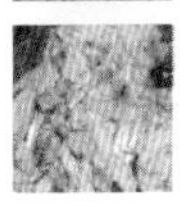	Radial clusters of soft, needle-like crystals found within cavities in basalt in northern Wisconsin	zeolite group, page 221

VIOLET TO PINK

If violet to pink and...	then try...
Soft cubic crystals, often occurring with calcite in limestone	fluorite, page 103
Very hard, glassy, six-sided pointed crystals or masses; most often found in northern Wisconsin	amethyst (quartz), page 197
Hard, pink, slender and elongated striated crystals embedded in granite	elbaite (tourmaline), page 217

METALLIC

If metallic and...	then try...
Crusts of brightly iridescent, colorful, metallic material from Wisconsin; often found on chalcocite	bornite, page 63
Blocky, dark metallic crystals from Wisconsin; often with a coating of brightly iridescent bornite	chalcocite, page 75
Fairly soft, brassy colored masses, veins or triangular crystals; often found with sphalerite in limestone	chalcopyrite, page 77
Soft, malleable reddish metal; often coated with malachite or chrysocolla	copper, page 93
Very heavy, metallic gray mineral occurring as cubic crystals or blocky masses	galena, page 127

Quick Identification Guide (continued)

(continued) **If metallic and...**	**then try...**
Metallic rust-brown crusts or masses, often with a botryoidal surface and a fibrous cross section	goethite, page 137
Very soft, highly malleable yellow metal found as tiny embedded flakes in quartz or limonite	gold, page 141
Common metallic mineral that is dark gray but is commonly stained reddish and may have a fibrous cross section	hematite, page 147
Tiny metallic black, slightly magnetic grains in dark rocks, particularly gabbro	ilmenite, page 149
Metallic black masses or octahedral crystals that bond strongly to a magnet	magnetite, page 159
Hard, brittle, brassy crystal groups with serrated edges or flat, striated, plate-like crystals; found in limestone	marcasite, page 165
Thin, hair-like, brassy needles; often found in clusters on calcite	millerite, page 171
Very soft, flexible, bluish gray metallic masses, veins or crusts that make a black mark on paper	molybdenite, page 173
Hard, brassy yellow to brown metallic cubes or faceted ball-like crystals or masses; pyrite is the most common brassy mineral	pyrite, page 189

If multicolored or banded and...	then try...
Very hard translucent masses containing ring-like banding within	agate, page 37
Very hard translucent banded masses or veins, often within limestone	sedimentary agate, page 39
Very hard, dense rock containing parallel layers of red jasper and metallic hematite or magnetite	banded iron formation, page 57
Rock consisting of smaller rounded or angular rocks cemented together	conglomerate or breccia, page 91
Round ball-like rock formations containing a hollow crystal-lined interior cavity	geodes, page 131
Generally hard, dense rocks containing coarse bands consisting of organized mineral layers	gneiss or schist, page 135
Abundant, coarse-grained rock with a mottled coloration; generally light colored with small dark spots	granite, page 143
Extremely coarse-grained rock with very large, readily identifiable crystals	pegmatite, page 181
Rocks with well-formed blocky crystals in otherwise fine-grained, dark rocks	porphyry, page 185

Quick Identification Guide (continued)

MULTICOLORED OR BANDED

(continued)	If multicolored or banded and...	then try...
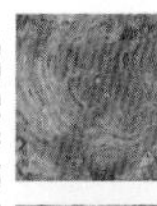	Rocks containing banded structures that are wave- or mushroom-like	stromatolites, page 119
	Hard pebbles that are green, orange or pink; often found along lakes or rivers	unakite, page 219

A final note about rock and mineral identification

When using this book to identify your rock and mineral discoveries, always remember that your specimens can (and likely will) differ greatly from those pictured. Rock and mineral identification isn't easy, and when a specimen is weathered or altered by external forces, it can appear completely different than it "should." So don't just compare your specimen to the photos in this book; learn the key characteristics of each rock or mineral and which traits are constant, such as hardness and crystal shape. With a basic understanding of quartz, for example, you'll be able to identify even the most poorly formed specimens. The same goes for other minerals, too.

Fibrous cross section of goethite mass
Botryoidal goethite
Associated
Radial fibrous
cross section
Fibrous mass
Botryoidal goethite
Limonite mass

Sample Page

HARDNESS: 7 **STREAK:** White

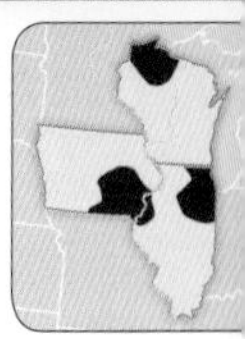

Primary Occurrence

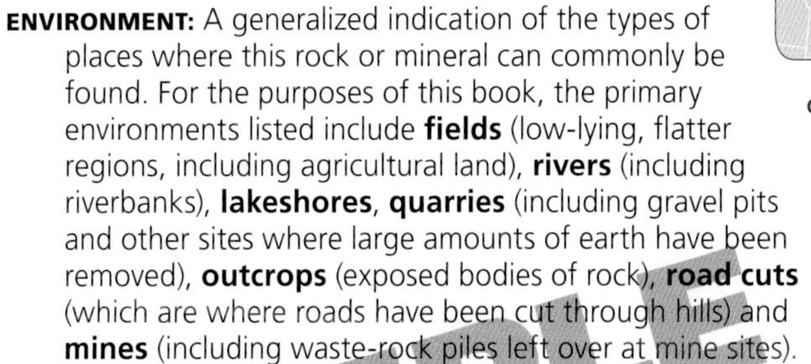

ENVIRONMENT: A generalized indication of the types of places where this rock or mineral can commonly be found. For the purposes of this book, the primary environments listed include **fields** (low-lying, flatter regions, including agricultural land), **rivers** (including riverbanks), **lakeshores**, **quarries** (including gravel pits and other sites where large amounts of earth have been removed), **outcrops** (exposed bodies of rock), **road cuts** (which are where roads have been cut through hills) and **mines** (including waste-rock piles left over at mine sites).

WHAT TO LOOK FOR: Common and characteristic identifying traits of the rock or mineral.

SIZE: The general size range of the rock or mineral. The listed sizes apply more to minerals and their crystals than to rocks, which typically form as enormous masses.

COLOR: The colors the rock or mineral commonly exhibits.

OCCURRENCE: The difficulty of finding this rock or mineral. "**Very common**" means the material takes almost no effort to find if you're in the right environment. "**Common**" means the material can be found with little effort. "**Uncommon**" means the material may take a good deal of hunting to find, and most minerals fall in this category. "**Rare**" means the material will take great lengths of research, time and energy to find, and "**very rare**" means the material is so scarce that you will be lucky to find even a trace of it.

NOTES: These are additional notes about the rock or mineral, including how it forms, how to identify it, how to distinguish it from similar minerals, and interesting facts about it.

WHERE TO LOOK: Here you'll find specific regions or towns where you should begin your search for the rock or mineral.

Polished agate
Natural agates
Agate in basalt
Waxy luster
Common beach-worn agates

Agates

HARDNESS: 6.5–7 **STREAK:** White

Primary Occurrence

ENVIRONMENT: Fields, rivers, lakeshores, quarries

WHAT TO LOOK FOR: Very hard, translucent, red or brown rounded masses of waxy material containing ring-like bands within

SIZE: Agates range from pebbles to specimens larger than a fist

COLOR: Multicolored; varies greatly, but banding is primarily red, brown, yellow, or white to gray; black is less common

OCCURRENCE: Uncommon

NOTES: As mysterious as they are popular, agates are banded gemstones that primarily consist of chalcedony (page 73), a compact microcrystalline form of quartz (page 195). Agates develop within cavities and cracks in rock, particularly within vesicles (gas bubbles) in volcanic rocks. Even though agates are known for their banding, it's not clear how these distinctive concentric bands form. Each band is a chalcedony shell enclosing smaller shells within it, not unlike the layers of an onion, and this unique structure is the primary way to distinguish agates from other forms of microcrystalline quartz, such as jasper (page 151) and chert (page 79). Jasper, chert, chalcedony and agates all exhibit conchoidal fracturing (when struck, circular cracks appear), extreme hardness, and a waxy luster, but jasper and chert are not translucent like chalcedony and agates, and chalcedony lacks agate-like banding. A large percentage of the agates in the region, especially those in Wisconsin, are actually Lake Superior agates and were transported southward by glaciers.

WHERE TO LOOK: Much of Wisconsin is rich with Lake Superior agates, particularly in the northern third of the state; look anywhere there is gravel. The Mississippi River is also lucrative; Lake Superior agates may be found anywhere along the river.

Iowa sedimentary agate in limestone
"Keswick agate"
"Nolte agate" in limestone
Agate layers formed on stromatolite fossil
Specimen courtesy of Phil Burgess

Agates, Sedimentary

HARDNESS: 6.5–7 **STREAK:** White

Primary Occurrence

ENVIRONMENT: Fields, rivers, lakeshores, quarries

WHAT TO LOOK FOR: Very hard, translucent masses of waxy, concentrically banded material within rock

SIZE: Sedimentary agates can range from tiny chips to boulders

COLOR: Multicolored; varies greatly, but banding is often red, orange, yellow to brown or white to gray

OCCURRENCE: Uncommon

NOTES: Most of the world's agates formed when hot, mineral-bearing volcanic water rose from deep within the earth and deposited chalcedony within a cavity, but Wisconsin, Iowa and Illinois each yield agates that formed from an entirely different process. These agates formed when cold surface water carrying dissolved minerals entered cavities in underlying rock. Called sedimentary or "coldwater" agates, they often formed within cracks or pockets in limestone (page 155), sometimes even within spaces made by stromatolites (page 119) and other fossils. Others developed as the cores of concretions (page 89) or geodes (page 131). But in nearly all sedimentary agates, specimens typically have rough, rocky and often "un-agate-like" exteriors. This can lead to confusion because chert and jasper can also form as concretions and may also have banding. But distinguishing between them isn't difficult, as the banding in sedimentary agates tends to be fine and sharply delineated (bands in chert are often "blurry") and often occurs with pockets of druzy quartz (page 195).

WHERE TO LOOK: Many sedimentary agates from the region are found in private quarries where you need to obtain permission to hunt. Nolte agates are found in quarries near Green Lake, Wisconsin. The Iowa towns of Keokuk and Keswick are each home to their own agate varieties.

Moss agates
Water-level agates
Close-up of moss agate
Rare Wisconsin thunder egg
Rocky exterior
Jagged agate-filled cavity

Agates, Varieties

HARDNESS: 6.5–7 **STREAK:** White

Primary Occurrence

ENVIRONMENT: Fields, rivers, lakeshores, quarries

WHAT TO LOOK FOR: Hard, translucent, red-to-brown masses of waxy banded material, sometimes with a round rock shell

SIZE: Agates can range from tiny fragments to fist-sized masses

COLOR: Multicolored; varies greatly, but banding is often red, brown to tan, yellow, white to gray, or sometimes black

OCCURRENCE: Uncommon to very rare

NOTES: Agates are such popular collectibles in part because they exhibit an amazing amount of variation. Sometimes a particular specimen's unique appearance is the result of variations in the agate's fundamental structure. This is best illustrated by water-level agates, which contain flat, parallel banding rather than the usual ring-like bands. Other unique variations seen within agates are the result of mineral inclusions, or other minerals that have grown within the agate. Moss agates are a very common example of this and contain worm- or ribbon-like growths of iron-bearing minerals. Both water-level and moss features are abundant within Lake Superior agate specimens. Both features can also be found in thunder eggs, which are among the rarest types of agates in the region. They consist of round, rocky masses containing a jagged, angular core of agate that may exhibit typical agate banding, water-level layers or moss inclusions, and formed in pockets made by expanding gases within freshly formed rock.

WHERE TO LOOK: Many interesting inclusions and variations can be seen in Lake Superior agates, which can be found in northern Wisconsin and in Mississippi River gravel bars in all three states. Wisconsin's thunder eggs are very rare, but the volcanic northern portions of the state, particularly Iron County, are good areas to look for them.

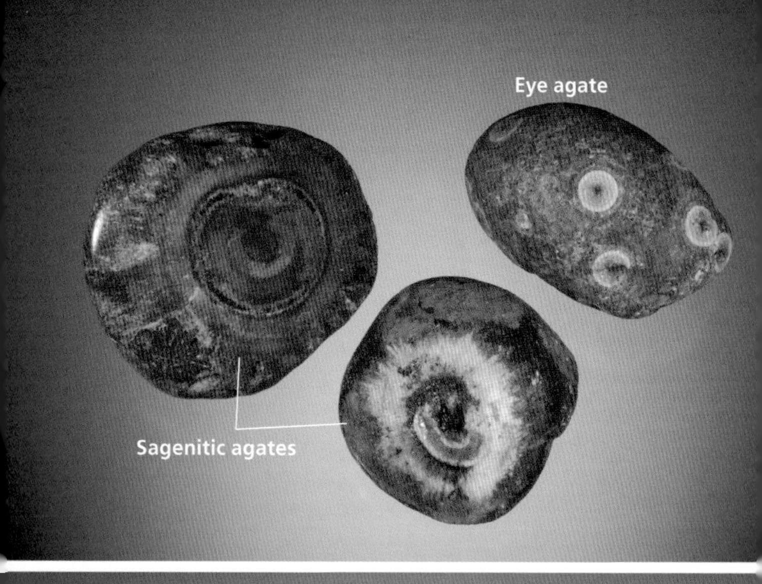
Eye agate
Sagenitic agates

Tube agates

Agates, Inclusions

HARDNESS: 6.5–7 **STREAK:** White

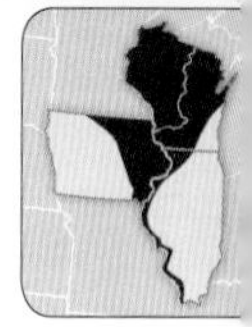

Primary Occurrence

ENVIRONMENT: Fields, rivers, lakeshores, quarries

WHAT TO LOOK FOR: Agates that contain unusual circular shapes

SIZE: Agates can range from tiny fragments to fist-sized masses

COLOR: Multicolored; varies greatly, but banding is often red, brown to tan, yellow, white to gray, or sometimes black

OCCURRENCE: Uncommon

NOTES: Agates exhibit a seemingly endless amount of variation, and this is particularly true when it comes to Lake Superior agates, which were spread throughout the region by the glaciers of past ice ages. Many agate variations are the result of mineral inclusions within the gemstones, some of which can greatly alter an agate's typical banding pattern. A fairly abundant example of this is commonly called a tube agate. Such agates literally contain tube-like structures. These tubes formed when needle-like crystals of minerals dissolved or were otherwise replaced by agate material, creating long, slender shapes and tunnel-like holes in agates. Another popular variety is even rarer; called a sagenitic agate, or a sagenite, it contains circular or fan-shaped radiating formations of needle-like structures. These formed when other minerals, especially goethite (page 137) or various zeolites (page 221), were replaced by chalcedony. In contrast, some agates contain circular features not created by inclusions. Known as eye agates, these specimens exhibit banded eye-like structures formed by the enlargement of normally microscopic chalcedony structures called spherulites. Sagenites and eyes can sometimes be confused; sagenitic growths generally aren't as perfectly circular as eyes.

WHERE TO LOOK: Sticking close to Lake Superior's shores, particularly near Superior, Wisconsin, is a good idea.

Hornblende crystals (black) in granite

Actinolite mass

Fibrous amphibole mineral (lustrous tan) on schist

Amphibole Group

HARDNESS: 5–6 **STREAK:** White

Primary Occurrence

ENVIRONMENT: Quarries, road cuts, outcrops, mines, rivers

WHAT TO LOOK FOR: Fairly hard, elongated dark crystals or masses embedded in rocks, often with a silky luster

SIZE: Most specimens are no larger than an inch

COLOR: Commonly gray to black; sometimes green to brown

OCCURRENCE: Common

NOTES: The amphiboles are a large and abundant group of minerals that are closely related to the pyroxenes (page 193). Like the pyroxenes, they are primarily "rock-building" minerals and are usually found as components of rocks. Grains, masses and crystals of various amphiboles are common in coarse-grained rocks like granite (page 143) and in metamorphic rocks, such as schists (page 135). Hornblende is one of the most common amphiboles, and it is common in granite where it can be seen as dark, lustrous crystals and masses that sometimes have a silky sheen. Other amphiboles, namely tremolite and actinolite, are present in some metamorphic rocks as elongated, fibrous masses, typically green to brown in color. Any dark needle-shaped crystals found embedded in coarse-grained rocks are often labeled as an amphibole, but differentiating between amphiboles and pyroxenes is difficult and requires additional research.

WHERE TO LOOK: Many fibrous amphiboles are associated with iron ore deposits, especially in northern Wisconsin where they can be found in schists neighboring iron mining areas. Closely inspecting any granite or gneiss from anywhere in the region will reveal grains of hornblende, and coarsely crystallized granite and granite-like rocks near Wausau may yield larger, more well-formed crystals.

Weathered anhydrite
Freshly exposed anhydrite

Anhydrite

HARDNESS: 3.5 **STREAK:** White

Primary Occurrence

ENVIRONMENT: Fields, quarries, outcrops

WHAT TO LOOK FOR: Gray, fairly soft, grainy mineral masses occurring with gypsum in sedimentary environments

SIZE: Anhydrite occurs massively and can be found in any size

COLOR: White to gray

OCCURRENCE: Rare

NOTES: Anhydrite is a mineral that formed primarily when ancient seas evaporated and left behind mineral deposits. In terms of composition, it is nearly identical to gypsum (page 145), and consists of calcium, sulfur and oxygen, but the key difference is that anhydrite lacks gypsum's water content. In fact, one of the ways anhydrite forms is by the dehydration of gypsum. Anhydrite is easy to mistake as a rock, especially when formed in large beds. Specimens are often gray, uncrystallized and rather mundane in appearance. Nonetheless, there are telling clues for collectors in the field, including its compact, even-textured nature and frequent occurrence with gypsum. Exposed beds may also exhibit cracks and smoothed surfaces that were dissolved by surface water. Distinguishing it from gypsum is rather simple, due to their great difference in hardness; it's more likely that it could be confused with massive calcite (page 65). Calcite, however, no matter what form it may take, is always more common and has rhombohedral cleavage, meaning that nearly any sample will break into a rhombohedron (a shape like a leaning cube) when carefully struck.

WHERE TO LOOK: Anhydrite is fairly rare in the region, found sparingly in only a few places. Exposed beds can be seen near Fort Dodge and other areas in central Iowa.

Finely crystallized aragonite with brown limonite

Specimen courtesy of Phil Burgess

Very fine glassy needle-like aragonite crystals

Specimen courtesy of Phil Burgess

Aragonite

HARDNESS: 3.5–4 **STREAK:** White

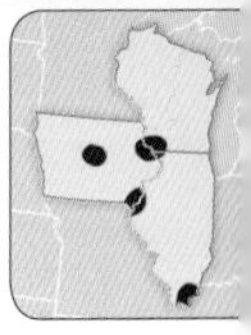

Primary Occurrence

ENVIRONMENT: Mines, outcrops, road cuts, fields

WHAT TO LOOK FOR: Light-colored, steeply pointed or needle-like crystals similar to calcite, but harder

SIZE: Crystals may be an inch or two long; masses may be several inches

COLOR: Colorless to white, gray, yellow to brown

OCCURRENCE: Rare

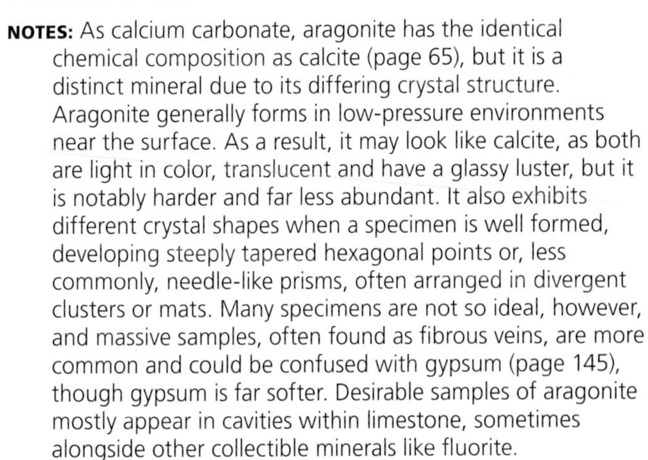

NOTES: As calcium carbonate, aragonite has the identical chemical composition as calcite (page 65), but it is a distinct mineral due to its differing crystal structure. Aragonite generally forms in low-pressure environments near the surface. As a result, it may look like calcite, as both are light in color, translucent and have a glassy luster, but it is notably harder and far less abundant. It also exhibits different crystal shapes when a specimen is well formed, developing steeply tapered hexagonal points or, less commonly, needle-like prisms, often arranged in divergent clusters or mats. Many specimens are not so ideal, however, and massive samples, often found as fibrous veins, are more common and could be confused with gypsum (page 145), though gypsum is far softer. Desirable samples of aragonite mostly appear in cavities within limestone, sometimes alongside other collectible minerals like fluorite.

WHERE TO LOOK: Aragonite isn't particularly common, but it may potentially be found in any sedimentary environments, particularly within cavities in limestone. Southern Illinois' Hardin County mines have produced amazing specimens alongside fluorite, but the area is largely off-limits. The area around Keokuk, Iowa, has also produced well-crystallized specimens.

Malachite (green)
Azurite veins in chalcopyrite
Limonite (orange)
Azurite veins (blue) in chalcopyrite with malachite (green)

Azurite

HARDNESS: 3.5–4 **STREAK:** Light blue

Primary Occurrence

ENVIRONMENT: Mines, quarries

WHAT TO LOOK FOR: Vivid blue veins or masses on copper or copper-bearing minerals, often alongside malachite

SIZE: Specimens tend to be smaller than a thumbnail

COLOR: Blue, often dark and vivid; described as "azure blue"

OCCURRENCE: Very rare in Wisconsin; not found in IA or IL

NOTES: When copper or copper-bearing minerals like chalcopyrite (page 77) weather, minerals like malachite (page 161), chrysocolla (page 83) and azurite typically develop nearby. Azurite is the rarest of these three weathering products, but it is perhaps the most coveted due to its typically bright, vivid and attractive blue coloration. Unfortunately, crystals larger than a fraction of a millimeter are not found in the region, but small masses and veins may rarely be recovered. Most specimens are glassy and lustrous, and when these characteristics are taken into consideration alongside its streak and hardness, identifying azurite isn't that difficult. Chrysocolla is somewhat similar, but it is more common and generally lighter and more green in color. Malachite is often present with azurite, due to their very similar composition and method of formation, but they are always colored differently. When looking for azurite in the region, magnification will be your biggest help; careful inspection of weathered copper-bearing minerals may reveal tiny veins or grains of azurite on their surfaces.

WHERE TO LOOK: You won't find azurite in Illinois or Iowa, but it may be found in Wisconsin anywhere there is copper, such as in Douglas County. It has also been found in Lafayette and Iowa Counties, particularly near Mineral Point where it occurs on chalcopyrite.

Barite crystal cluster

Glassy luster

Intergrown crystals

Cluster of fine parallel crystals from a geode

Barite (Baryte)

HARDNESS: 3–3.5 **STREAK:** White

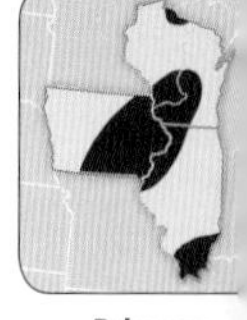

Primary Occurrence

ENVIRONMENT: Quarries, mines, road cuts, fields

WHAT TO LOOK FOR: Light-colored blocky or bladed crystals that feel very heavy for their size, often grown in more-or-less parallel groups that exhibit step-like features on their faces

SIZE: Barite crystals are typically smaller than an inch or two, but crystal clusters may rarely grow to fist-sized masses

COLOR: Gray to white, tan to yellow or brown, less commonly red

OCCURRENCE: Uncommon

NOTES: Barite (spelled "baryte" outside the U.S.) is the most abundant barium-bearing mineral and is a fairly common find in sedimentary rocks, particularly in lead- and zinc-bearing limestones like those found in the region. When well crystallized, it takes the form of glassy elongated, blocky prisms with wedge-shaped tips and generally square cross sections, but flat, thin, plate- or blade-like crystals grown in layered, parallel groups are also common. In either form, barite crystals often exhibit small step-like features on their surface, and thanks to the high density of barium, any specimen of barite will feel heavy for its size (which is unusual for a glassy mineral). Other forms of barite include irregular masses and fibrous "balls," which will be harder to identify. Barite often occurs with galena and sphalerite; it is harder than calcite and softer than fluorite.

WHERE TO LOOK: Geodes from the Keokuk, Iowa, area may rarely contain barite blades, but it is mines in the Buffalo area that have produced some of the finest specimens in the region. Many nice specimens have originated from Hardin County, Illinois; other good specimens are found near Mount Carroll. Dozens of sites in Wisconsin produce barite as well, especially in the galena-rich limestone of Lafayette County.

Rough basalt
Mineral-lined vesicles
Texture detail
Mineral-filled vesicles
Water-worn basalt

Basalt

HARDNESS: 5–6 **STREAK:** N/A

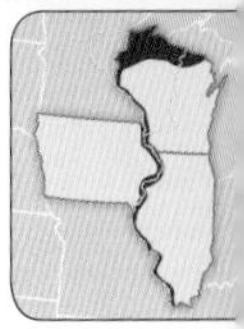

Primary Occurrence

ENVIRONMENT: Lakeshore, rivers, outcrops, quarries

WHAT TO LOOK FOR: Dark gray to black, dense, fine-grained rock, often containing many vesicles (gas bubbles)

SIZE: As a rock, basalt can be found in any size

COLOR: Gray to black, sometimes with a greenish tint; red to brown when weathered

OCCURRENCE: Common in Wisconsin; uncommon elsewhere

NOTES: Basalt, one of the planet's most prevalent rocks, forms as a direct result of volcanic activity, forming when lava (molten rock) erupts onto the earth's surface. During the Midcontinent Rift, in which North America was trying to split in two, enormous amounts of lava rose into the Lake Superior region, hardening rapidly in the cool atmosphere. In contrast with rocks that cooled deep within the ground where the earth's interior insulated them and kept them semi-solid for eons, basalt cooled and solidified so quickly that virtually none of the minerals contained within it were able to crystallize to a visible size. And because basalt is comprised primarily of plagioclase feldspar, olivine, pyroxenes, amphiboles and magnetite—all generally dark minerals—the result is a nearly black, very fine-grained rock with an evenly colored appearance. Its rapid cooling also trapped gases within it, which are visible as vesicles, or round bubble-like cavities. Rhyolite (page 201) can look very similar, but is slightly harder, has slightly larger mineral grains, is lighter in color and is more common in north-central Wisconsin.

WHERE TO LOOK: Basalt is most abundant in northern Wisconsin in any county bordering Lake Superior, particularly Ashland and Douglas counties. But the glaciers pushed basalt into Iowa and Illinois as well, where you'll find it in river gravel.

Banded iron formation
Specimens courtesy of Phil Burgess
Texture detail
Banded iron formation
Jasper layers (red)
Iron ore layers (black)
Highly weathered layers

Banded Iron Formation

HARDNESS: 6–7 **STREAK:** N/A

Primary Occurrence

ENVIRONMENT: Mines, outcrops, quarries

WHAT TO LOOK FOR: Very hard masses containing differently colored layers, often with a metallic "glitter" and red jasper

SIZE: Banded iron formations form as enormous masses and can be found in any size

COLOR: Multicolored; varies greatly, but primarily layered in colors of brown to bright red, tan to gray, and metallic black

OCCURRENCE: Uncommon

NOTES: Banded iron formation, or BIF for short, is one of the Lake Superior region's most interesting rocks. Found throughout the famous iron ranges that surround the lake, all of the region's BIFs formed over a billion years ago when iron- and silica-rich sediments settled to the bottom of oxygen-poor seas. While it's not exactly clear how BIF formed, it's thought that ancient life, particularly cyanobacteria, a type of blue-green algae evident today as stromatolites (page 119), created oxygen via photosynthesis. This oxygen combined with iron particles, forming hematite (page 147) and magnetite (page 159), the prominent iron minerals in BIF. As time progressed, these sediments cemented into a very hard rock containing large amounts of chert (page 79) and jasper (page 151) layered between the bands of iron ores. There is little you will confuse with BIF; the opaque, parallel, partially metallic and often magnetic layering is distinctive.

WHERE TO LOOK: The Gogebic Iron Range in Ashland and Iron counties in northern Wisconsin are the only places you'll find BIF in the region (except for glacially deposited pieces elsewhere). Highway 77 in Iron County follows the Gogebic Range and is dotted with old iron mine sites; the Hurley area is surrounded by iron mineral-bearing outcrops.

White Benstonite crystals on calcite
Stacked crystal habit
Calcite
Fine intergrown rhombohedral crystals
Calcite

Benstonite

HARDNESS: 3–4 **STREAK:** White

Primary Occurrence

ENVIRONMENT: Mines

WHAT TO LOOK FOR: White rhombohedral crystals with calcite or fluorite; found in southern Illinois

SIZE: Individual crystals are smaller than a thumbnail, but clusters may be several inches

COLOR: White, tan to pale yellow; rarely colorless

OCCURRENCE: Very rare in Illinois; not found in IA or WI

NOTES: Benstonite is an extremely rare mineral in the region, found only in the world famous Minerva No. 1 Mine in southern Illinois' legendary fluorite-producing Hardin County. A complex strontium- and barium-bearing carbonate mineral, Benstonite is typically found as well-formed rhombohedral crystals (shaped like a leaning or skewed cube). Its crystals grow clustered together in stacks or crusts on the surfaces of other minerals, particularly fluorite (page 127), calcite (page 65) and barite (page 53). Almost always white in color and opaque, you might confuse it with calcite or barite because of its low hardness, but at the Minerva Mine, calcite generally occurs as steeply pointed crystals and barite specimens are found in thin bladed crystals. By contrast, Benstonite has thick, blocky, intergrown and stacked rhombohedrons.

WHERE TO LOOK: In this region, Benstonite is only found in Illinois, particularly the Cave-in-Rock area in Hardin County where just one mine produced the mineral. The finest crystals in the world were once recovered there, but Benstonite is about as rare as it gets in the upper Midwest, and unless you obtain permission to enter the off-limits mine and hire a professional to guide you, you won't find it yourself anywhere but on a rock shop's shelf (and it won't be cheap!).

Quartz-lined geode stained brown
by dried bitumen
Specimen courtesy of Kevin Ponzio
Bitumen-coated
quartz crystal
Moist asphaltum (brown) on fluorite

Bitumen/Asphaltum

HARDNESS: N/A **STREAK:** N/A

Primary Occurrence

ENVIRONMENT: Fields, quarries, road cuts, mines

WHAT TO LOOK FOR: Black or brown sticky or greasy fluid, or blob-like masses on or within other minerals, such as fluorite

SIZE: Concentrations of bitumen may be up to fist-sized or larger

COLOR: Black to brown

OCCURRENCE: Rare

NOTES: The upper Midwest, particularly Illinois, is rich with hydrocarbons (oil-compounds consisting of hydrogen and carbon) derived from ancient organic material. Hydrocarbons are not minerals, but they often appear in mineralized sedimentary areas, occurring alongside minerals like calcite, quartz, fluorite, and even within geodes. Bitumen is the name given to semi-solid hydrocarbons found in these environments, and it is the result when a more fluid hydrocarbon called asphaltum has largely dried out. This material was transported through the host rock, which was folded and otherwise subjected to pressure, forcing the hydrocarbons upwards and into pockets and cavities where minerals happened to be forming. As a result, bitumen and asphaltum often appear as blob-like droplets or smeared coatings on rock or mineral surfaces. Often rich in sulfur, hydrocarbons can have a pungent smell—this is often the first clue they are present—and they are typically sticky to the touch, making them easy to identify.

WHERE TO LOOK: Many of the famous but now off-limits mines in Hardin County, Illinois, produced coatings and blobs of bitumen on fluorite and other fine minerals. Some very rare geodes in central and eastern Iowa can also be broken open to reveal a pool of bitumen or asphaltum within.

Pale-pink crust on chalcocite
Bluish bornite coating on chalcocite

Bornite

HARDNESS: 3 **STREAK:** Dark gray

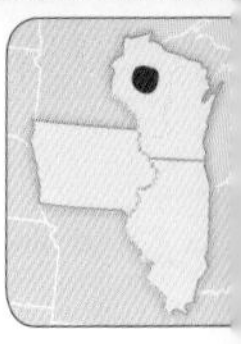

Primary Occurrence

ENVIRONMENT: Mines

WHAT TO LOOK FOR: A soft, rare, metallic, dark bronze mineral, often with an iridescent surface coating

SIZE: Masses may be up to several inches, but crystals are thumbnail-sized or smaller

COLOR: Brown-bronze; typically tarnished and multicolored pink, purple, blue and greenish

OCCURRENCE: Very rare in Wisconsin; not found in IA or IL

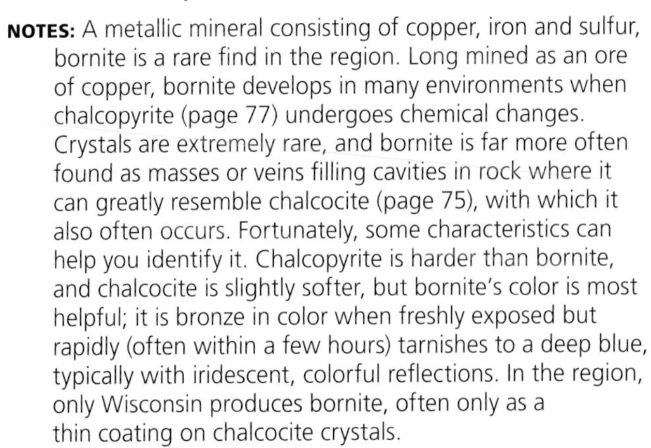

NOTES: A metallic mineral consisting of copper, iron and sulfur, bornite is a rare find in the region. Long mined as an ore of copper, bornite develops in many environments when chalcopyrite (page 77) undergoes chemical changes. Crystals are extremely rare, and bornite is far more often found as masses or veins filling cavities in rock where it can greatly resemble chalcocite (page 75), with which it also often occurs. Fortunately, some characteristics can help you identify it. Chalcopyrite is harder than bornite, and chalcocite is slightly softer, but bornite's color is most helpful; it is bronze in color when freshly exposed but rapidly (often within a few hours) tarnishes to a deep blue, typically with iridescent, colorful reflections. In the region, only Wisconsin produces bornite, often only as a thin coating on chalcocite crystals.

WHERE TO LOOK: The famous Flambeau Mine in Rusk County, Wisconsin, has produced legendary specimens of chalcocite coated by beautifully iridescent microscopic bornite crystals, as well as masses and veins of bornite. Unfortunately, the mine is off-limits, and there are few, if any, opportunities for collectors to find their own bornite in the region. It does, however, turn up for sale in old collections and in shops.

Fine calcite crystal

Older generation of brown calcite

Barite

Steeply pointed crystals

Rhombohedral crystal fragment

Quartz

Calcite crystals

Geode containing quartz and calcite

Calcite

HARDNESS: 3 **STREAK:** White

Primary Occurrence

ENVIRONMENT: All environments

WHAT TO LOOK FOR: Light-colored, six-sided pointed crystals or blocky masses or veins that are easily scratched with a U.S. nickel (coin); breaks into a "leaning cube" shape

SIZE: Crystals and masses can measure up to several inches

COLOR: Colorless to white most common; often yellow to brown

OCCURRENCE: Very common

NOTES: Calcite, like quartz (page 195), is so abundant that it is one of the very first minerals every collector should learn to recognize. Composed of calcium carbonite, calcite can form in nearly any geological environment, but is particularly prevalent within the region's sedimentary rocks. It has several hundred known crystal forms, but it most commonly appears as elongated, steeply angled hexagonal (six-sided) points, which are sometimes pointed at both ends. It also frequently develops as rhombohedrons (a shape resembling a leaning cube). Irregular crusts, veins or masses of calcite are the most common find, but calcite is easy to identify in any form. Generally light colored and fairly soft, calcite is easily scratched by a U.S. nickel, distinguishing it from quartz. It also has rhombohedral cleavage, meaning that any specimen will break into perfect rhombohedrons. Calcite effervesces (fizzes) in acids; test it with undiluted vinegar.

WHERE TO LOOK: Spectacular specimens come from limestone quarries near Buffalo in Scott County, Iowa, and elsewhere along the Mississippi River. Geodes from limestone near Hamilton, Illinois, contain crystals, and large specimens are found in the famous Hardin County mines. In Wisconsin, some of the best specimens have come from mines and outcrops throughout southern Lafayette County.

Countless tiny rhombohedral calcite crystals on limestone

Botryoidal calcite

Calcite embedded in limestone

Calcite-filled cracks in septarium

Calcite (continued)

HARDNESS: 3 **STREAK:** White

Primary Occurrence

ENVIRONMENT: All environments

WHAT TO LOOK FOR: Light-colored blocky crystals, masses or veins that are easily scratched with a U.S. nickel; breaks into a "leaning cube" shape

SIZE: Crystals and masses can measure up to several inches

COLOR: Colorless to white most common; often yellow to brown

OCCURRENCE: Very common

NOTES: Calcite is so abundant in the region that many visually diverse varieties exist, especially on and in limestone (which is itself composed largely of calcite). In fact, when in the field in Illinois, Iowa or southern Wisconsin, nearly any limestone exposure will yield calcite in one form or another, whether it is crystalline or massive. Rhombohedral (shaped like a leaning cube) crystals embedded in limestone are reflective and visible when the stone is rotated in bright light, and checking the weathered cracks and spaces in limestone may reveal fairly hard, botryoidal (lumpy, grape-like) crusts of calcite that were redeposited after it had been dissolved from the surrounding rock. Any pocket or cavity in limestone may also contain countless tiny rhombohedral crystals, which are sometimes only apparent in bright light when the interior of a cavity "sparkles." And finally, calcite is found filling in cracks and gaps in almost any kind of rock, including septariums, which are clay-rich concretions (page 89) that long ago "dried out," leaving a network of cracks which later filled in with other minerals, particularly calcite.

WHERE TO LOOK: Most of southern Wisconsin and all of Iowa and Illinois contain so much limestone that inspecting any boulder will reveal calcite-filled pockets or crusts.

Mass of intergrown celestine crystals
Single crystal
Mass of crystals
Finely formed crystals

Celestine

HARDNESS: 3–3.5 **STREAK:** White

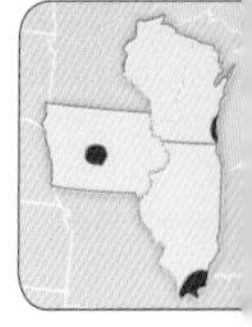

Primary Occurrence

ENVIRONMENT: Mines, quarries

WHAT TO LOOK FOR: Soft, fibrous masses or layers in limestone; also glassy, blocky, pale blue crystals, often with fluorite

SIZE: Masses may be up to several inches; crystals are typically smaller than an inch

COLOR: Colorless to white or gray common, sometimes pale blue

OCCURRENCE: Very rare

NOTES: Celestine may be the most common strontium-bearing mineral, but it's a rare find in Wisconsin, Illinois and Iowa. Closely related to barite (page 53) in structure, celestine occurs in sedimentary rocks, particularly within limestone or in beds with gypsum (page 145). Most of the time it is found as grainy or fibrous masses embedded in its host rock; identifying such celestine specimens can be difficult due to their nondescript nature, and if it weren't for its hardness and rarity (you won't likely stumble upon it), it could be easily mistaken for gypsum. Crystallized specimens are far more impressive but even rarer. They take the form of elongated blocky prisms, have a high luster, and are tucked in cavities or grow in clusters. Celestine clusters look similar to those of barite; the best way to identify them is by subtle differences in crystal shape. Celestine's tend to be more complex and angular on their tips; barite is more blade-like in shape.

WHERE TO LOOK: Celestine is rare all over the region; quarries near Milwaukee, Wisconsin, have yielded small amounts of crystals, and gypsum mines in Iowa, particularly Webster County, have produced massive examples. But by far the best specimens in the region originated from the famous Hardin County mines in far-southern Illinois; unfortunately the mines are off-limits to collecting.

Cerussite coating (tan) on galena
Purer, better-crystallized cerussite (white)
Cerussite coating (tan) on galena

Cerussite

HARDNESS: 3–3.5 **STREAK:** White

Primary Occurrence

ENVIRONMENT: Quarries, road cuts, mines, outcrops

WHAT TO LOOK FOR: Small light-colored, brightly lustrous crystals or dull, dusty coatings on galena

SIZE: Crystals tend to be no larger than ¼ inch; coatings may be up to several inches

COLOR: Colorless to white if crystallized; typically gray to tan

OCCURRENCE: Uncommon

NOTES: Cerussite is a mineral that can be very easily overlooked, thanks to its relatively uninteresting appearance. Related to aragonite (page 49), cerussite is a lead-bearing mineral (wash after handling) most abundant as a weathering product of galena (page 127). It forms as galena crystals' outer surfaces break down. Therefore it is most common as a dull, dusty white or gray coating on the surface of an otherwise metallic and lustrous galena specimen. In these cases, it is best identified simply by its association with galena. In rare cases, larger crystals of cerussite can be found; they occur as stubby prisms with low-pointed tips and are sometimes as lustrous as glass. (Such finds were more common decades ago when lead mining still dominated much of the region.) Well-crystallized finds can be confused with other minerals, such as aragonite or calcite (page 65), but their rarity and association with galena will help you identify them if you are lucky enough to recover a specimen. Regardless of appearance, it can be safely assumed that some amount of cerussite is present on virtually any galena specimen.

WHERE TO LOOK: Southern Wisconsin, particularly the Mineral Point region, was once the site of a major lead-mining industry; any piece of galena found in the area today will likely be coated in tan cerussite.

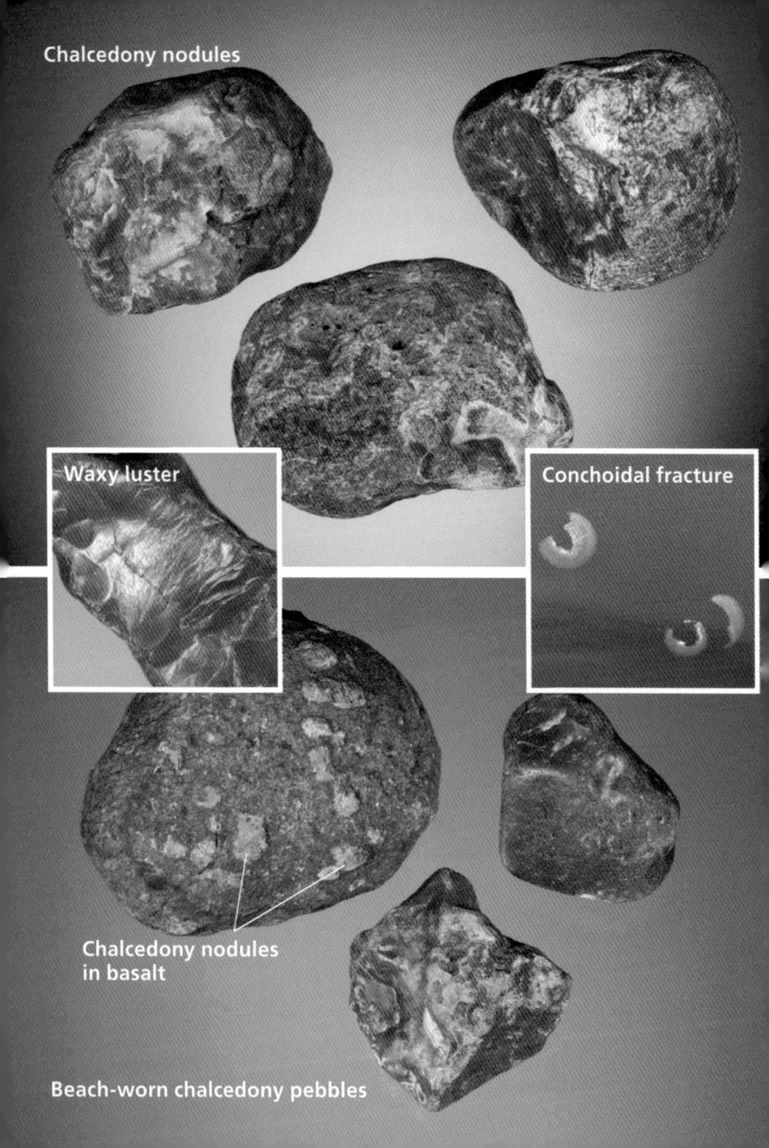
Chalcedony nodules
Waxy luster
Conchoidal fracture
Chalcedony nodules
in basalt
Beach-worn chalcedony pebbles

Chalcedony

HARDNESS: 6.5–7 **STREAK:** White

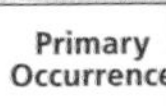

Primary Occurrence

ENVIRONMENT: Lakeshores, rivers, fields, quarries

WHAT TO LOOK FOR: Very hard, waxy masses of translucent material with a varied, mottled coloration

SIZE: Chalcedony can form in a wide range of sizes, though it is typically found as pieces smaller than your fist

COLOR: Varies; primarily white to gray when pure; red to brown is common; also yellow to green, and often multicolored

OCCURRENCE: Common

NOTES: Chalcedony is a microcrystalline form of quartz (page 195), which means that it forms in indistinctly shaped masses that consist of countless microscopic quartz crystals. Chalcedony is famous as the material of which agates are formed (page 37). This means that chalcedony is dense and very hard, so hard that weathered pieces appear polished with a waxy luster. It also shares the traits of quartz itself, such as its high hardness, translucency and conchoidal fracture (when struck, circular cracks appear). Chalcedony can take many shapes, from nodules (round formations) to vein-filling masses or botryoidal (lumpy, grape-like) growths within geodes. It is often light colored, though red and brown hues are common, the result of iron-rich impurities. Jasper (page 151) and chert (page 79) are easily confused with chalcedony, but both are opaque; chalcedony is virtually always translucent, but you might need a bright light or a thin fragment to observe translucency.

WHERE TO LOOK: You'll find chalcedony in glacial till and river gravel anywhere. The banks of the Mississippi River in the Chain of Rocks area in Illinois, near St. Louis, is rich with specimens. Geodes from Illinois and particularly near Keswick, Iowa, may contain botryoidal chalcedony. And in Wisconsin, the shores of Lake Superior are full of samples.

Crude intergrown chalcocite crystals coated in orange bornite
Iridescent coating
Crude, parallel crystals
Mass of chalcocite (black and iridescent) with chalcopyrite (brassy)

Chalcocite

HARDNESS: 2.5–3 **STREAK:** Shiny black

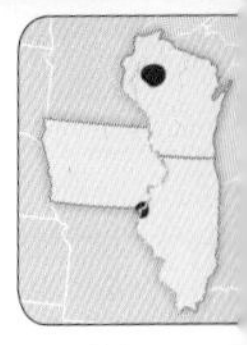

Primary Occurrence

ENVIRONMENT: Mines, outcrops, quarries

WHAT TO LOOK FOR: A dark, gray, soft metallic mineral occurring as veins or small elongated crystals; often with malachite

SIZE: Crystals are rarely larger than ¼ inch (some extraordinary examples measure several inches); masses may be fist-sized

COLOR: Lead-gray to metallic black; rarely with iridescent coating

OCCURRENCE: Very rare

NOTES: Chalcocite is composed of copper and sulfur, an easily separated mixture that makes it an important ore of copper. Most commonly occurring as indistinct veins or masses within various kinds of rock, chalcocite is metallic and generally dark gray in color. It is soft, brittle, and can typically be found with some amount of malachite (page 161), which is the result of weathering. But while most specimens are rather mundane, the region has also produced very rare world-class chalcocite crystals, particularly from Rusk County, Wisconsin. When ideally formed, crystals are elongated and rectangular with a tapered, pointed tip, but many are formed poorly or are twinned (intergrown) in such a way that they appear to be hexagonal (six-sided) rather than elongated. The famous Wisconsin specimens are frequently coated in a beautiful iridescent coating of microscopically crystallized bornite (page 63). Non-crystallized chalcocite could be confused with galena (page 127), but galena is a bit softer and will cleave, or break, into perfect cubes.

WHERE TO LOOK: Geodes from the areas around Keokuk, Iowa and Hamilton, Illinois, rarely contain tiny crystals, but it is the now-closed Flambeau Mine near Ladysmith in Rusk County, Wisconsin, that has produced some of the finest in the world. Polk County, Wisconsin, also contains several collecting sites.

Chalcopyrite crystals (brassy yellow) on sphalerite (black)
Crystals with dolomite
Crystal with calcite
Chalcopyrite veins in limestone

Chalcopyrite

HARDNESS: 3.5–4 **STREAK:** Greenish black

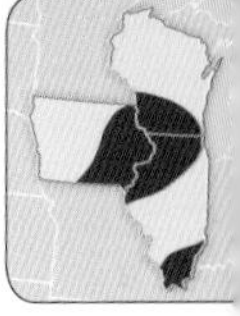
Primary Occurrence

ENVIRONMENT: Mines, quarries, outcrops, road cuts

WHAT TO LOOK FOR: Brittle, fairly soft, brass-yellow masses, veins or triangular crystals; often found with sphalerite

SIZE: Chalcopyrite crystals are rarely larger than your thumbnail, but masses or veins may be several inches in size

COLOR: Brass-yellow to golden yellow, metallic brown; often with an orange surface coating or blue to purple iridescence

OCCURRENCE: Common in WI and IL; uncommon in IA

NOTES: Consisting of copper, iron and sulfur, chalcopyrite (cal-co-pyrite) is an attractive mineral that primarily formed in the region as a result of chemical reactions in limestone. Found throughout all three states, it is always brassy in color, similar to pyrite (page 189), but it is typically more yellow and softer than pyrite. The exception is when chalcopyrite is weathered, in which case it is often more orange in color, resulting from a coating of limonite (page 157). Other weathered specimens may exhibit a bluish to purplish iridescent surface coating, similar to that of bornite (page 63), though bornite is slightly softer and is rarer. And though irregular masses are most abundant, crystals are not rare and most often appear as tetrahedrons (triangular pyramids) that are easy to identify by their triangular shapes; this shape helps distinguish it from marcasite (page 165). Finally, where you find sphalerite (page 211), you'll likely find chalcopyrite.

WHERE TO LOOK: The southwestern corner of Wisconsin has produced fine specimens from pockets in limestone, especially in Iowa County near Mineral Point. In Illinois, Hardin County's famous mines have produced nice crystals with fluorite; rare, but fine, specimens are recovered from within geodes in Iowa, particularly near Keokuk.

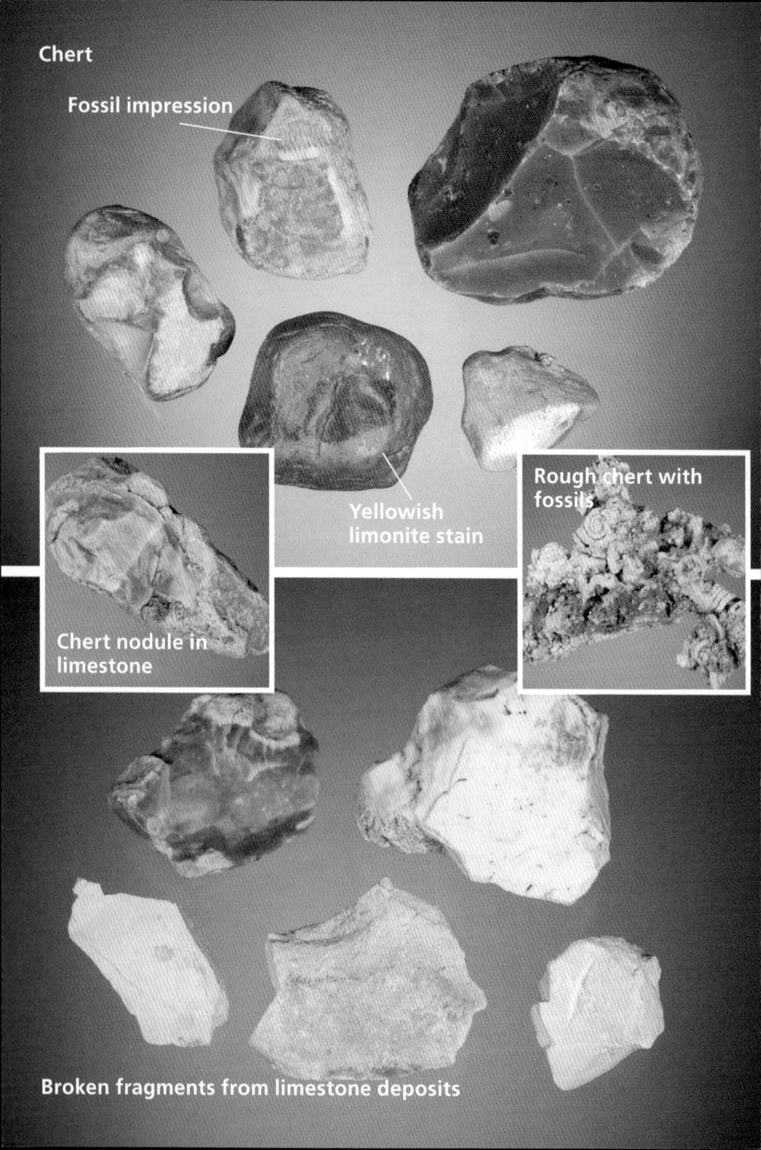
Chert
Fossil impression
Yellowish limonite stain
Rough chert with fossils
Chert nodule in limestone
Broken fragments from limestone deposits

Chert

HARDNESS: ~7 **STREAK:** N/A

Primary Occurrence

ENVIRONMENT: All environments

WHAT TO LOOK FOR: Very hard opaque masses, often gray to brown; waxy and rounded when found in rivers or gravel

SIZE: As a rock, chert can range from pebbles to boulders in size

COLOR: White to gray, black, tan to yellow, brown

OCCURRENCE: Very common

NOTES: Chert is an extremely hard, dense and opaque sedimentary rock composed almost entirely of compacted microscopic grains of quartz (page 195). Though it can form in several ways, in this region it is commonly found both as pockets and veins within limestone or as large layers or beds between other sedimentary rocks. In the former example, it developed within the limestone when silica (quartz material) solutions seeped into the rock and concentrated, sometimes centered around a fossil. Many concretions (page 89) formed this way. In the latter method of formation, the sedimentary layers of chert developed when the remains of diatoms (a type of algae that grows rigid skeletons made of silica) settled onto seafloors in thick beds that later solidified. Chert will likely be one of the hardest rocks you'll find, matched only by chalcedony (page 73), which is more translucent, and quartzite (page 199), which is grainier and more glassy in appearance. Chert is typically dull and rough, but it is so hard that weathering can nearly polish it, making rounded, water-worn specimens appear waxy and smooth. It also shares quartz's conchoidal fracture (when struck, circular cracks appear), which will help identify it.

WHERE TO LOOK: Chert is abundant throughout the region; river gravel will contain countless pebbles. Rough chert containing fossils can be found in far southwestern Wisconsin.

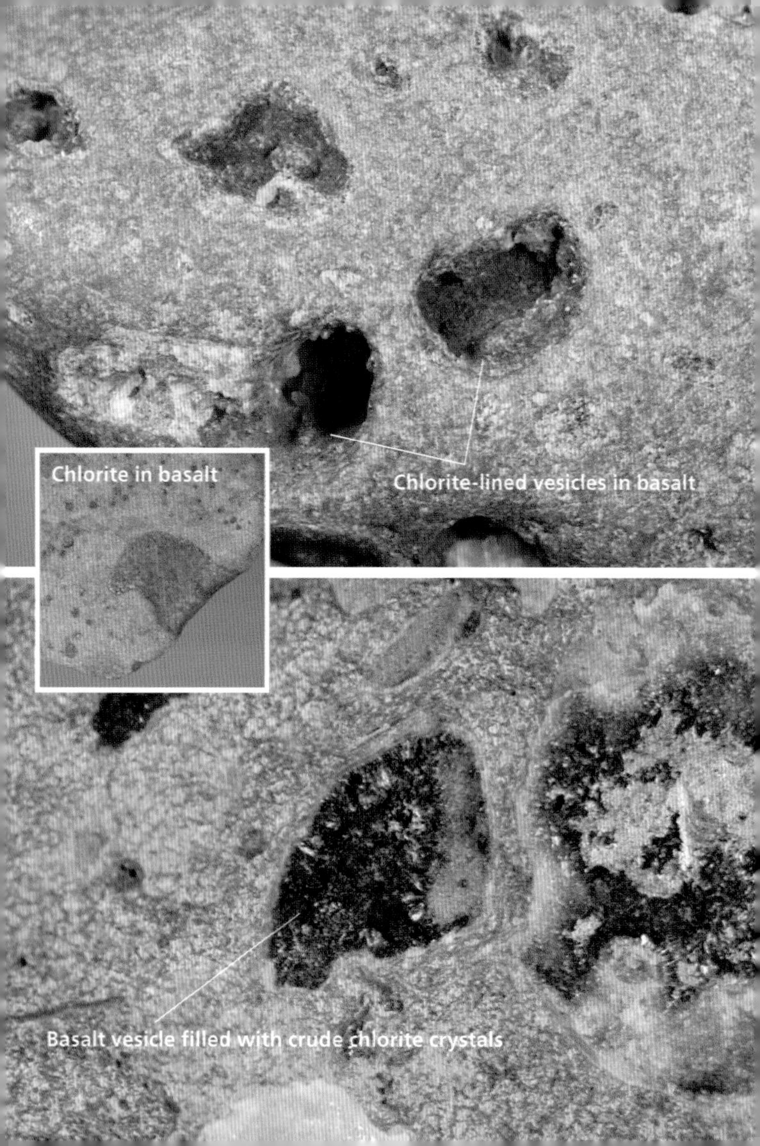
Chlorite in basalt
Chlorite-lined vesicles in basalt
Basalt vesicle filled with crude chlorite crystals

Chlorite Group

HARDNESS: 2–2.5 **STREAK:** Colorless to green

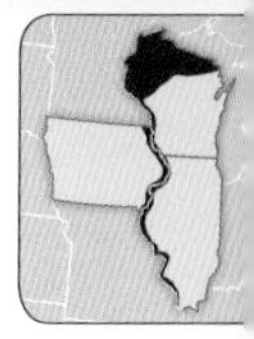

Primary Occurrence

ENVIRONMENT: All environments

WHAT TO LOOK FOR: Soft, dark, "greasy"-looking material lining cavities in rocks; tiny grains in greenish colored rocks

SIZE: Individual crystals are tiny and generally microscopic, but crusts, veins or masses can measure up to an inch or two

COLOR: Green to dark green, occasionally brown to black

OCCURRENCE: Common in Wisconsin; uncommon in IA and IL

NOTES: If you've ever picked up a piece of slate or schist with a green tint, it was likely colored by chlorite group minerals. There are several chlorite minerals, but because they are all indistinguishable from each other outside of a lab, we simply call them all "chlorite." In this region, you'll mostly find chlorite in Wisconsin where it is a common constituent of metamorphic rocks like schist (page 135). It formed when amphiboles (page 45), olivine (page 179) and pyroxenes (page 193) in metamorphic rocks were chemically altered. It also turns up in vesicles (gas bubbles) in basalt in the Lake Superior region. In all cases, you typically won't find chlorite crystals, so you'll have to rely on other traits to identify this often mundane and easily overlooked mineral. Chlorite is very soft and generally dark green in color with a "greasy" luster, which makes masses of it fairly easy to identify. In metamorphic rocks, chlorite is often only present as tiny grains, so an overall greenish tint may be your only clue.

WHERE TO LOOK: You may find greenish chlorite-bearing rocks all over the region in glacial till, but the only decent specimens you'll find are in northern Wisconsin. In Douglas County, chlorite is very prominent in basalt vesicles, and slates, schists and gneisses in Bayfield, Ashland and Iron Counties contain chlorite as a primary constituent.

Chrysocolla
Copper
Chrysocolla with azurite (dark blue)
Specimen courtesy of Rob Carlson
Chrysocolla on copper
Chrysocolla
Azurite (dark blue)

Chrysocolla

HARDNESS: 2–4 **STREAK:** White to pale blue

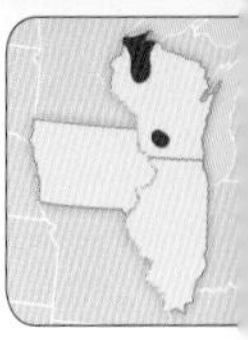

Primary Occurrence

ENVIRONMENT: Mines, outcrops

WHAT TO LOOK FOR: Soft, crumbly bluish green crusts or masses on or near copper, often with malachite

SIZE: Chrysocolla can be found as small masses up to an inch

COLOR: Bluish green, pale blue

OCCURRENCE: Rare in Wisconsin; not found in IA or IL

NOTES: Chrysocolla is a rewarding mineral to collect as it is generally vividly colored and quite easy to identify. It is a weathering product of copper-bearing minerals and formed when those minerals were affected by oxygen, groundwater and other environmental factors. It is therefore always associated with copper-related minerals, including chalcopyrite (page 77), chalcocite (page 75) and even native copper (page 93) itself. It typically is found as a crust on the surface of or within the nearby rock, and it is often intergrown with malachite (page 161). In general, chrysocolla has little recognizable structure. It is typically found as a dusty coating or crumbly crust on its parent mineral, but botryoidal (lumpy, grape-like) masses or tiny delicate fibers can rarely be found. While rare in the region, identifying chrysocolla is a simple matter of noting its color—generally always blue, sometimes pale or with a greenish tint—in conjunction with its nondescript structure, low hardness, crumbly nature, and association with copper-bearing minerals.

WHERE TO LOOK: You'll only find chrysocolla in Wisconsin, and only in locations known for copper, such as the famous copper-rich mining district in Rusk County, near Ladysmith. Northern Wisconsin yields small amounts of copper, particularly near Lake Superior and in Douglas County; look near Gordon and Solon Springs.

Clay "pool" in quartz-lined geode

Clay-rich shale

Iron-rich rocky clay

Clay-coated quartz crystals in geode

Clay Minerals

HARDNESS: ~1–2 **STREAK:** N/A

Primary Occurrence

ENVIRONMENT: All environments

WHAT TO LOOK FOR: Very soft masses of fine-grained material that have a chalky or earthy texture and easily crumble

SIZE: Individual clay mineral crystals are microscopic; beds or masses of clay can be enormous

COLOR: Off-white to gray when pure; more often brown to red

OCCURRENCE: Very common

NOTES: Clay, the sticky mud often encountered along rivers, is something everyone is familiar with, but it actually consists of several minerals. The clay minerals are an assortment of chemically different but structurally similar minerals that develop as microscopic flat, plate-like crystals. These crystals are arranged into stacks too small to see without a powerful microscope. Clay exposed along rivers and lakes tends to be quite impure, containing clay minerals but also tiny grains of rocks and other minerals as well as organic detritus. Sedimentary rock outcrops often reveal layers of purer clay that formed when aluminum-bearing minerals chemically weathered. In our region, several clay minerals are present, including illite, montmorillonite, kaolinite, dickite and halloysite. Of these, illite is perhaps the most common, making up the bulk of most of the region's clay. While identifying individual clay minerals is difficult, determining if you've found clay minerals is easy. Clays are soft and crumbly when dry, but malleable when wet. Finally, clays are common in the region's famous geodes and are visible as soft, white coatings or "pools" on or between crystals.

WHERE TO LOOK: Clays are ubiquitous throughout all three states and are found as layers in, or between, sedimentary rocks. Geodes from Keokuk, Iowa, may yield clay within them.

Anthracite fragments

Coal

HARDNESS: <3 **STREAK:** N/A

Primary Occurrence

ENVIRONMENT: Mines, quarries, fields, outcrops

WHAT TO LOOK FOR: Black, shiny, lightweight material found in sedimentary rocks or in quarries

SIZE: Masses of coal can occur in any size

COLOR: Dark gray to black, brown-black

OCCURRENCE: Common in IA and IL; uncommon in WI

NOTES: Long used as a combustible fossil fuel, coal is a lightweight black material consisting of fossil plant matter that never fully decayed. When aquatic plants die in acidic water that lacks oxygen, decomposition can halt, causing the plant matter to build up. Over time, overlying sediment can compress that material, concentrating the plants' carbon into beds. As the pressure increases, coal begins to form, first as lignite, a soft, woody variety of coal, and eventually resulting in anthracite, the final form of coal, which is created only under high-pressure. Anthracite coal has undergone enough change that it is actually considered a rock. Thanks to the upper Midwest's aquatic past, coal deposits are common and it is mined throughout the region. Identification is simple due to coal's soft and lightweight nature, its very shiny luster and its combustibility. If the region's coal had been subjected to further increasing pressure, purer forms of carbon could have resulted, including graphite and diamond (unfortunately for collectors, this didn't happen).

WHERE TO LOOK: Many coal mines have operated in south-central Iowa, and though the mines are off-limits, coal may be found throughout that part of the state. Illinois holds extensive coal reserves, which are found throughout most of the state, particularly in the south and west. In Wisconsin, only the southernmost portion of the state yields coal.

Mudstone concretion
Bulbous forms
Mudstone concretion
Sandstone concretion
Sandstone concretion
Specimen courtesy of Phil Burgess

Concretions

HARDNESS: N/A **STREAK:** N/A

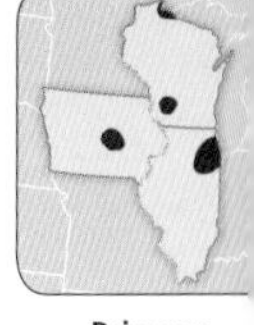
Primary Occurrence

ENVIRONMENT: Rivers, outcrops, fields, lakeshores

WHAT TO LOOK FOR: Conspicuously round, ovoid, and/or lumpy-shaped rocks found within sedimentary rock environments

SIZE: Concretions can be up to a foot or more in size

COLOR: Brown to tan, yellow, gray; often depends on host rock

OCCURRENCE: Uncommon

NOTES: Concretions are rather peculiar rounded rock formations. They are sometimes so mundane that you'd walk over them, but others are so unusual and enigmatic that you can't help but notice. They form in two ways. Some develop within beds of sedimentary rock, such as sandstone and mudstone, when mineral-bearing solutions periodically collect around a central point. Others form when organic material decays and releases carbon and other compounds that react with the surrounding sediment. This causes a sphere of mineral-rich rock to form around the organic nucleus. Both processes result in round or strangely lumpy, grainy-textured rocks that are often freed from their host rock by weathering. Unlike geodes (see page 131), concretions are not hollow, and they are generally harder than the rock in which they formed but consist of a similar material. For example, concretions formed in sandstone are themselves made of sandstone. Differentiating between a concretion and a stone that was rounded by weathering can take some research.

WHERE TO LOOK: Some of the most striking concretions have been found near Lake Superior around Ashland and Bayfield in northern Wisconsin. Sandstone concretions are common near Boscobel, Wisconsin, and along the Wisconsin River. The Mazon Creek area near Morris, Illinois, produces siderite concretions that formed around fossil matter.

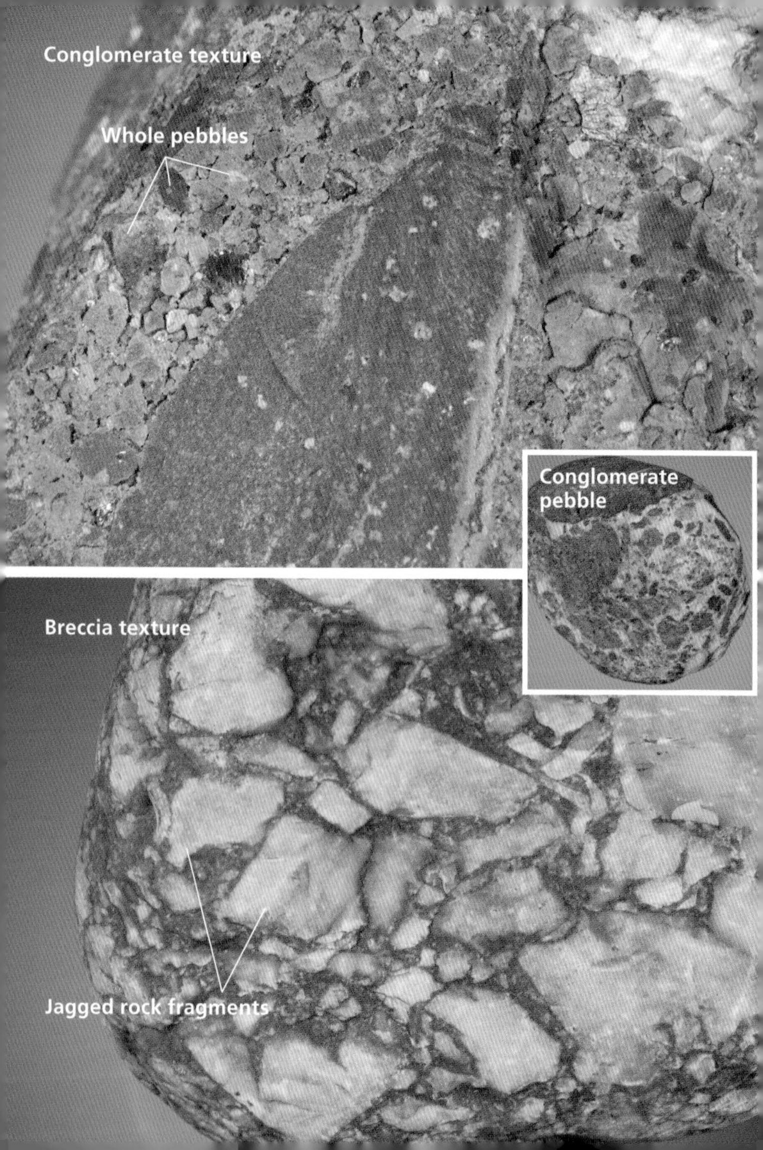
Conglomerate texture
Whole pebbles
Conglomerate pebble
Breccia texture
Jagged rock fragments

Conglomerate/Breccia

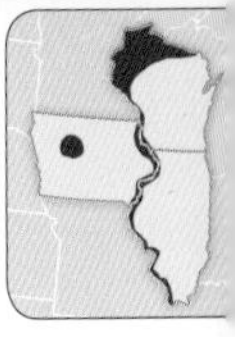

Primary Occurrence

HARDNESS: N/A **STREAK:** N/A

ENVIRONMENT: Rivers, quarries, road cuts, outcrops

WHAT TO LOOK FOR: Rocks that appear to be made up of many smaller rocks that have been cemented together

SIZE: Conglomerate and breccia can be found in any size

COLOR: Varies greatly; mottled and multicolored

OCCURRENCE: Uncommon

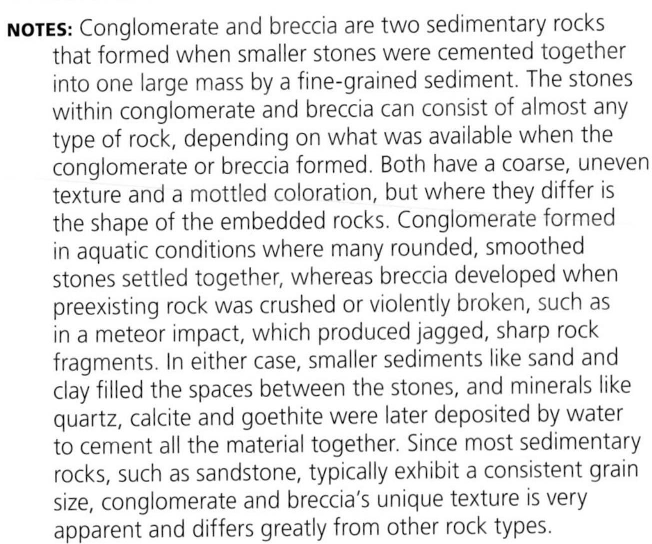

NOTES: Conglomerate and breccia are two sedimentary rocks that formed when smaller stones were cemented together into one large mass by a fine-grained sediment. The stones within conglomerate and breccia can consist of almost any type of rock, depending on what was available when the conglomerate or breccia formed. Both have a coarse, uneven texture and a mottled coloration, but where they differ is the shape of the embedded rocks. Conglomerate formed in aquatic conditions where many rounded, smoothed stones settled together, whereas breccia developed when preexisting rock was crushed or violently broken, such as in a meteor impact, which produced jagged, sharp rock fragments. In either case, smaller sediments like sand and clay filled the spaces between the stones, and minerals like quartz, calcite and goethite were later deposited by water to cement all the material together. Since most sedimentary rocks, such as sandstone, typically exhibit a consistent grain size, conglomerate and breccia's unique texture is very apparent and differs greatly from other rock types.

WHERE TO LOOK: Brecciated rocks can be found in the vicinity of the Manson Impact Structure in Pocahontas and Calhoun counties in Iowa, especially near Manson itself. Conglomerate can be found in the Lake Superior region in northern Wisconsin, and anywhere within glacial till in Illinois.

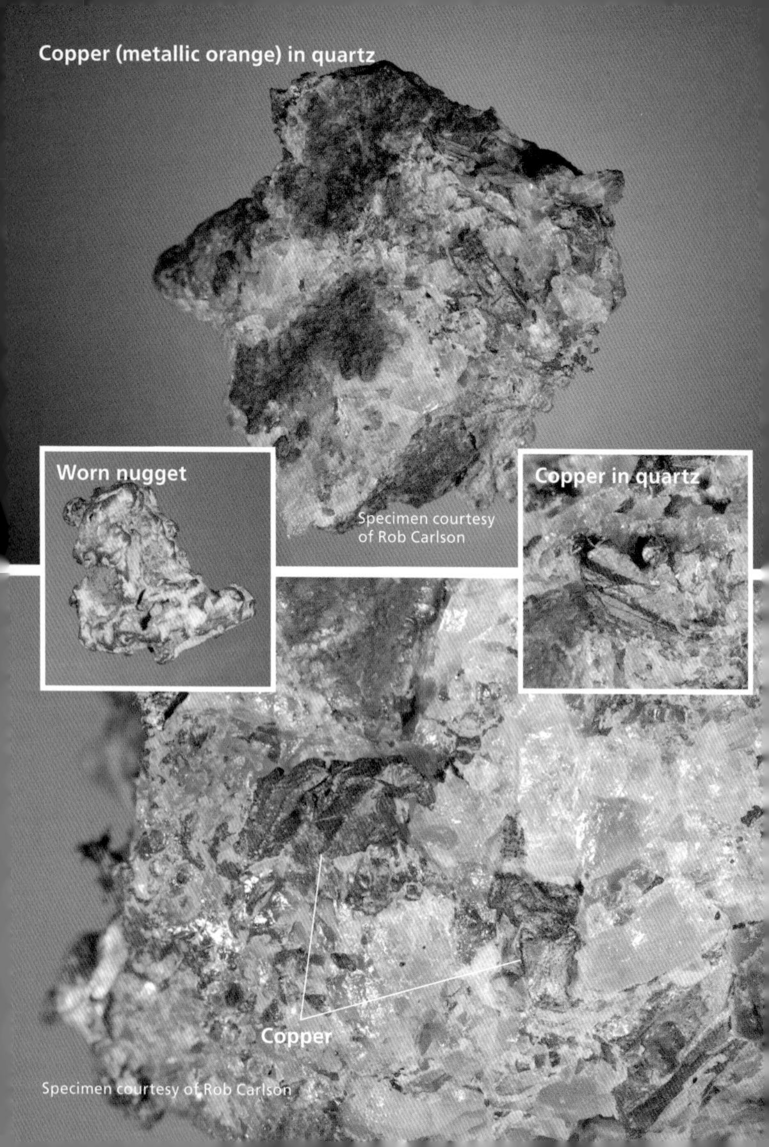
Copper (metallic orange) in quartz
Worn nugget
Specimen courtesy of Rob Carlson
Copper in quartz
Copper
Specimen courtesy of Rob Carlson

Copper

HARDNESS: 2.5–3 **STREAK:** Metallic orange-red

Primary Occurrence

ENVIRONMENT: Mines, outcrops, rivers, quarries

WHAT TO LOOK FOR: Reddish-orange metal, often in veins or cavities within dark rocks and coated in greenish material

SIZE: Masses of copper can be enormous, but you'll rarely find specimens larger than your fist

COLOR: Metallic orange-red when freshly exposed; a dark metallic brown, reddish, or blue to green when weathered

OCCURRENCE: Uncommon in Wisconsin; very rare in IA and IL

NOTES: Like gold, copper is a native element, which means that, unlike most minerals, it consists of a single element and therefore has very uniform and constant traits, such as the metallic reddish-orange color for which copper is known. In natural specimens, however, that hallmark color is often hidden beneath coatings of green malachite (page 161) and blue chrysocolla (page 83), which formed when the outer surfaces of a piece of copper weathered, combining with other elements. In these three states, you'll really only find copper in northern Wisconsin, where it was deposited in cavities within basalt that formed during the Midcontinent Rift, which is the same way Michigan's famed copper deposits developed. Copper can also be found further south in Wisconsin, thanks to the glaciers, which rolled the copper into rounded nuggets. In all cases, copper is easy to identify because of its color, low hardness and high malleability.

WHERE TO LOOK: With the exception of some very rare nuggets pushed into Illinois gravel pits by glaciers, copper is only found in Wisconsin. It is particularly common in Douglas County near Lake Superior, where it can be found in cavities in basalt, especially near South Range, Gordon and along the Amnicon River. River gravel may also contain masses.

Water-worn diabase

Lighter-colored spots

Diabase texture detail

Gabbro texture detail

Coarse texture

Gabbro fragments

Diabase/Gabbro

HARDNESS: >5.5 **STREAK:** N/A

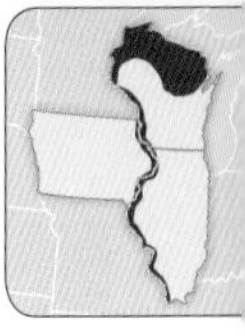

Primary Occurrence

ENVIRONMENT: Lakeshore, rivers, fields, outcrops

WHAT TO LOOK FOR: Dark-colored, coarsely grained rocks containing many glassy mineral grains and rectangular crystals

SIZE: As rocks, diabase and gabbro can be found in most any size

COLOR: Mottled; predominantly black to greenish gray with lighter-colored spots; browner when weathered

OCCURRENCE: Common in Wisconsin; uncommon elsewhere

NOTES: During the Midcontinent Rift event, northern Wisconsin was flooded with lava, which cooled to form dark igneous rocks. Basalt (page 55) was the primary rock that formed, but diabase and gabbro, two closely related rocks, are prevalent as well. When the molten rock that formed basalt cooled on the earth's surface, it solidified very rapidly, resulting in a rock with very fine mineral grains. If the same molten rock had cooled within the earth, the minerals would have grown to much larger sizes. Diabase and gabbro have very similar mineral compositions to basalt, but differ in grain size because of the depth at which they formed within the earth. Gabbro cooled at the lowest depth and contains large, jagged, angular crystals of plagioclase feldspar, olivine, pyroxenes, and other dark minerals. It tends to have a mottled, dark greenish tint and brightly reflective glassy mineral grains. Diabase cooled at a shallower depth and has a grain size between that of gabbro and basalt. It looks like basalt, but has larger and more visible light-colored grains.

WHERE TO LOOK: Thanks to the glaciers, pebbles of these rocks may be found in glacial till all over the region. With that said, they are primarily a Wisconsin find. All Wisconsin counties that border Lake Superior, particularly Ashland County, have extensive outcrops of igneous rocks.

Dolomite crystals (orange) in geode
Fine crystals
Curved crystals
Chalcopyrite crystals
Well-formed dolomite crystals

Dolomite Group

HARDNESS: 3.5–4 **STREAK:** White

Primary Occurrence

ENVIRONMENT: Quarries, outcrops, mines, fields, rivers, road cuts

WHAT TO LOOK FOR: Blocky, rhombohedral (shaped like a leaning cube) tan crystals, typically with pearly, curved surfaces

SIZE: Individual crystals are rarely larger than ½ inch, but crystal clusters or masses may be up to several inches in size

COLOR: White to tan or brown, more rarely orange or dark gray

OCCURRENCE: Common

NOTES: Dolomite is a rather common mineral that most often forms within cavities in sedimentary rocks, primarily limestone. A close cousin to calcite (page 65), dolomite forms as light-colored rhombohedral crystals, which resemble a leaning or skewed cube, and it also often exhibits curved or rounded crystal faces with a soft, pearly luster. The most practical way to tell dolomite from calcite, however, is by its hardness, as dolomite is harder. You'll most often find dolomite crystals in vugs (irregular cavities) and veins in limestone alongside minerals like chalcopyrite and sphalerite. Ankerite, a member of the dolomite group, is a very closely related mineral with ample iron content in its composition, but it is rarer and impossible to distinguish without lab equipment. Finally, a variety of dolomite-rich limestone, called dolostone (or, confusingly, "dolomite" in older texts) is prevalent in the region. It is more weather resistant but visually the same as limestone.

WHERE TO LOOK: Much of the rock lining the Mississippi River in southwest Wisconsin and northeast Iowa is dolostone. Fine dolomite crystals are frequent in geodes from Hamilton, Illinois, and Keokuk, Iowa, and any limestone formations have the potential to produce specimens from within cavities.

Quartz
Epidote crystals
Epidote with quartz (white) in basalt
Beach-worn pebble
Epidote coating on basalt

Epidote

HARDNESS: 6–7 **STREAK:** Colorless to gray

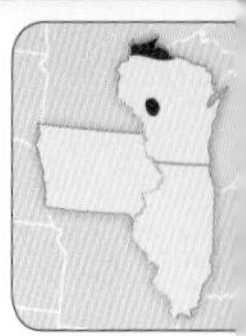

Primary Occurrence

ENVIRONMENT: Lakeshores, rivers, outcrops, roadcuts, quarries

WHAT TO LOOK FOR: Hard, yellow-green elongated crystals, crusts or masses in rock, particularly basalt or gabbro

SIZE: Individual crystals are generally smaller than ¼ inch

COLOR: Green to yellow-green, sometimes dark green

OCCURRENCE: Uncommon in Wisconsin; very rare in IA and IL

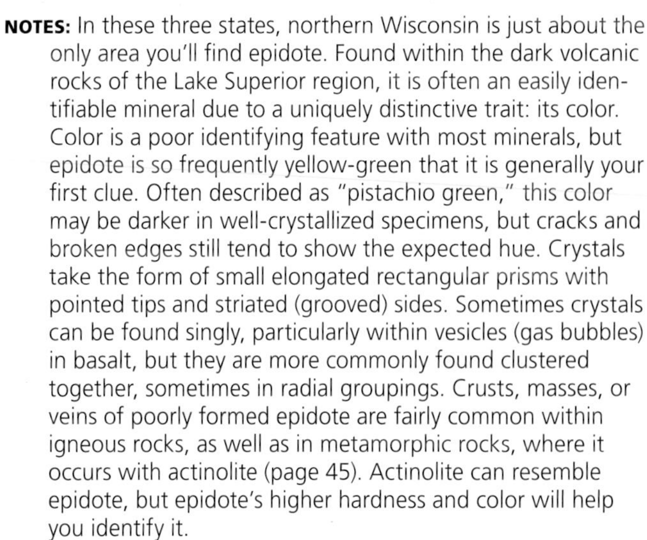

NOTES: In these three states, northern Wisconsin is just about the only area you'll find epidote. Found within the dark volcanic rocks of the Lake Superior region, it is often an easily identifiable mineral due to a uniquely distinctive trait: its color. Color is a poor identifying feature with most minerals, but epidote is so frequently yellow-green that it is generally your first clue. Often described as "pistachio green," this color may be darker in well-crystallized specimens, but cracks and broken edges still tend to show the expected hue. Crystals take the form of small elongated rectangular prisms with pointed tips and striated (grooved) sides. Sometimes crystals can be found singly, particularly within vesicles (gas bubbles) in basalt, but they are more commonly found clustered together, sometimes in radial groupings. Crusts, masses, or veins of poorly formed epidote are fairly common within igneous rocks, as well as in metamorphic rocks, where it occurs with actinolite (page 45). Actinolite can resemble epidote, but epidote's higher hardness and color will help you identify it.

WHERE TO LOOK: With the exception of epidote carried southward by glaciers, you'll only find epidote in Wisconsin. Basalt outcrops near Lake Superior in Douglas and Ashland counties and near Grandview in Bayfield County are some of the primary epidote areas.

Cluster of fine microcline crystals
K-spar in granite
Feldspars (tan/pink) in granite
Cleaved fragments
Water-worn feldspar pebbles

Feldspar Group

HARDNESS: 6–6.5 **STREAK:** White

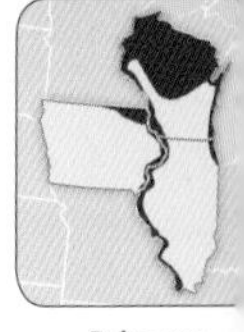

Primary Occurrence

ENVIRONMENT: All environments

WHAT TO LOOK FOR: Abundant, hard, light-colored masses embedded in granite, or blocky, angular crystals in cavities

SIZE: Varies greatly; masses or grains range from tiny flecks to an inch or more in size; crystals can be up to fist-sized

COLOR: White to gray, pink to orange, tan to brown

OCCURRENCE: Very common

NOTES: Present in some form within nearly every type of rock, feldspars are one of the major building blocks of our planet, comprising nearly 60 percent of the earth's crust. But "feldspar" is just a general term for the entire group, which is actually divided into two subgroups: potassium feldspars ("K-spars," for short), such as the very common orthoclase and microcline, and the plagioclase feldspars, such as albite and anorthite. K-spars are abundant in light-colored rocks like granite (page 143) and pegmatite (page 181) as embedded, light-colored, opaque blocky masses or grains, or more rarely as large crystals with low-angled tips. Plagioclase feldspars, on the other hand, are often only found in dark rocks as glassy grains. When embedded, all feldspars can be difficult to identify; to identify them, look for finds that are opaque, have a high hardness and exhibit rectangular or blocky shapes. Many specimens also exhibit a pearly sheen or internal schiller, which is distinctive, as is feldspar's blocky cleavage (when broken, square step-like breaks appear).

WHERE TO LOOK: Crystals are best found in Wisconsin; in Douglas County, you can find small orange crystals in basalt cavities, and large crystals have been found in granite quarries near Rib Mountain in Marathon County. Elsewhere, loose pebbles are found in glacial till or river gravel, as are granite stones.

Very fine fluorite crystal cluster
Cubic crystals
Broken mass
Cubic crystal
Calcite
Brown fluorite crystals
Same specimen fluorescing whitish under short-wave ultraviolet light

Fluorite

HARDNESS: 4 **STREAK:** White

Primary Occurrence

ENVIRONMENT: Quarries, mines, outcrops

WHAT TO LOOK FOR: Glassy, translucent masses or cubic crystals that can be scratched by a knife but not by a U.S. nickel

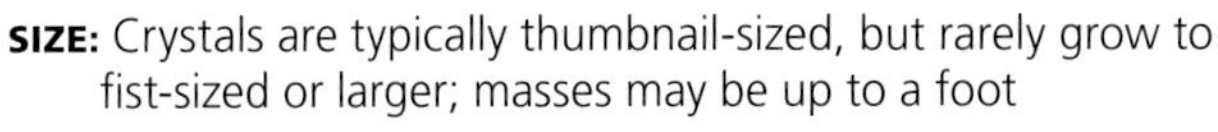

SIZE: Crystals are typically thumbnail-sized, but rarely grow to fist-sized or larger; masses may be up to a foot

COLOR: Yellow to brown, purple, colorless; rarely blue

OCCURRENCE: Uncommon

NOTES: The upper Midwest is known for its world-class fluorite specimens, particularly those from southern Illinois, which grace museum collections all over the globe. Fluorite, the most abundant fluorine-bearing mineral, is a relatively common occurrence in the type of limestone formation that is prevalent in the region, and it is popular among collectors, thanks to its often vivid yellow-to-purple color and charming cubic crystals. Crystals are typically intergrown with each other, sometimes forming beautiful complex clusters of cubes, and they often occur with numerous other minerals, including calcite, galena, chalcopyrite, sphalerite, and even bitumen. In Iowa and Illinois, it is found primarily in pockets or veins within limestone, but in northern Wisconsin, it can rarely be found in granite. When colored, identification isn't difficult, but colorless or pale specimens may resemble calcite (page 65), though calcite is notably softer.

WHERE TO LOOK: Hardin County, Illinois, is the source of some of the world's finest fluorites, but collecting is off-limits and specimens command high prices. In eastern Iowa, particularly Black Hawk County, outcrops and quarries have yielded fine specimens, many of which are wonderfully fluorescent, and in Wisconsin, the granite quarries near Wausau in Marathon County have produced small crystals.

Quartz with evidence of stromatolites
Fossil snail shells
Specimens courtesy of Phil Burgess
Stromatolite
Snail shell fragment on chert
Reed fossil in concretion
Fern fossil in concretion

Fossils

HARDNESS: N/A **STREAK:** N/A

Primary Occurrence

ENVIRONMENT: Fields, rivers, outcrops, road cuts, quarries, lakeshores

WHAT TO LOOK FOR: Rocks, especially limestone, containing structures and impressions that resemble living things

SIZE: Most fossils are smaller than your palm, but some may be up to several feet

COLOR: Tan to brown or gray; colored like the surrounding rock

OCCURRENCE: Uncommon

NOTES: Iowa, Illinois and the southern half of Wisconsin are home to large amounts of fossils, including some that have proved extremely important to our understanding of ancient organisms. It therefore goes without saying that fossil collectors love this region, as there are ample opportunities to find specimens. Fossils form when organic matter, such as a tooth, shell or leaf, is buried in sediment, particularly underwater, where it cannot decay normally. Over millions of years, minerals from the surrounding sediment and rock react with the once-living material and replace its cells, turning it into a mineral formation. This process occurred more often with the hard parts of animals, since they were less delicate and therefore more apt to survive fossilization, but both animal and plant fossils exist in the region. They can be found in shale and sandstone, but limestone (page 155) is the most common environment for fossils, as it formed at the bottom of ancient seas. Not coincidentally, most of the region's fossils are of aquatic animals.

WHERE TO LOOK: Much of Iowa, especially the northeastern portion, is rich with fossils of all kinds, as is southwestern Wisconsin. Near the Mississippi River, any rock outcrop and river gravel may yield specimens.

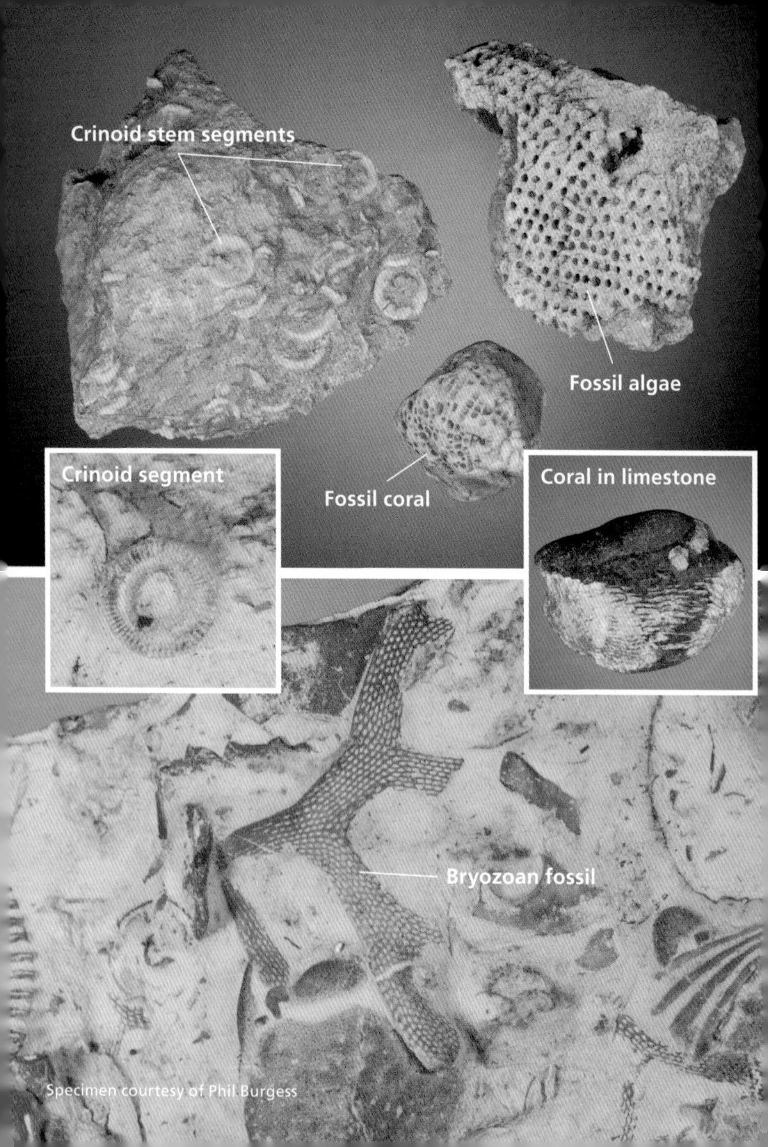

Crinoid stem segments

Fossil algae

Crinoid segment

Fossil coral

Coral in limestone

Bryozoan fossil

Specimen courtesy of Phil Burgess

Fossils, Aquatic

HARDNESS: N/A **STREAK:** N/A

Primary Occurrence

ENVIRONMENT: Fields, rivers, outcrops, road cuts, quarries

WHAT TO LOOK FOR: Rocks, especially limestone, containing round or branching structures, often with a gauze-like texture

SIZE: Aquatic fossils can be quite large, up to a foot or larger

COLOR: Tan to brown or gray; colored like the surrounding rock

OCCURRENCE: Uncommon

NOTES: Numerous shallow seas have inundated the upper Midwest over the eons, leaving behind copious amounts of sedimentary rocks, particularly limestone (page 155) and chert (page 79). Within them are the fossil remains of ancient aquatic life. The most abundant fossils consist of colonial animals, such as coral and bryozoans, and segments of anchored animals, such as crinoids. Other common fossils include impressions of various species of algae and traces of bacterial growths, namely stromatolites (page 119). Identifying such fossils can be tricky because some of the forms are subtle and are hard to spot, while others may seem strange and difficult to define. Corals tend to have a segmented structure, often with a gauze-like appearance and texture, and may be rounded in shape. Bryozoans, while similar to coral, are often seen in smaller, more delicate, branching forms. Crinoid "stems" are found as segmented tubular structures, or as loose round disks when separated. Finally, algae can be very difficult to identify, but many appear as rounded or blob-like structures marked with "pin-holes."

WHERE TO LOOK: Eastern Iowa and southwestern Wisconsin are rich with fossils of all kinds. Near the Mississippi River, any rock outcrop and river gravel may yield specimens where the water has cut through sedimentary rock layers.

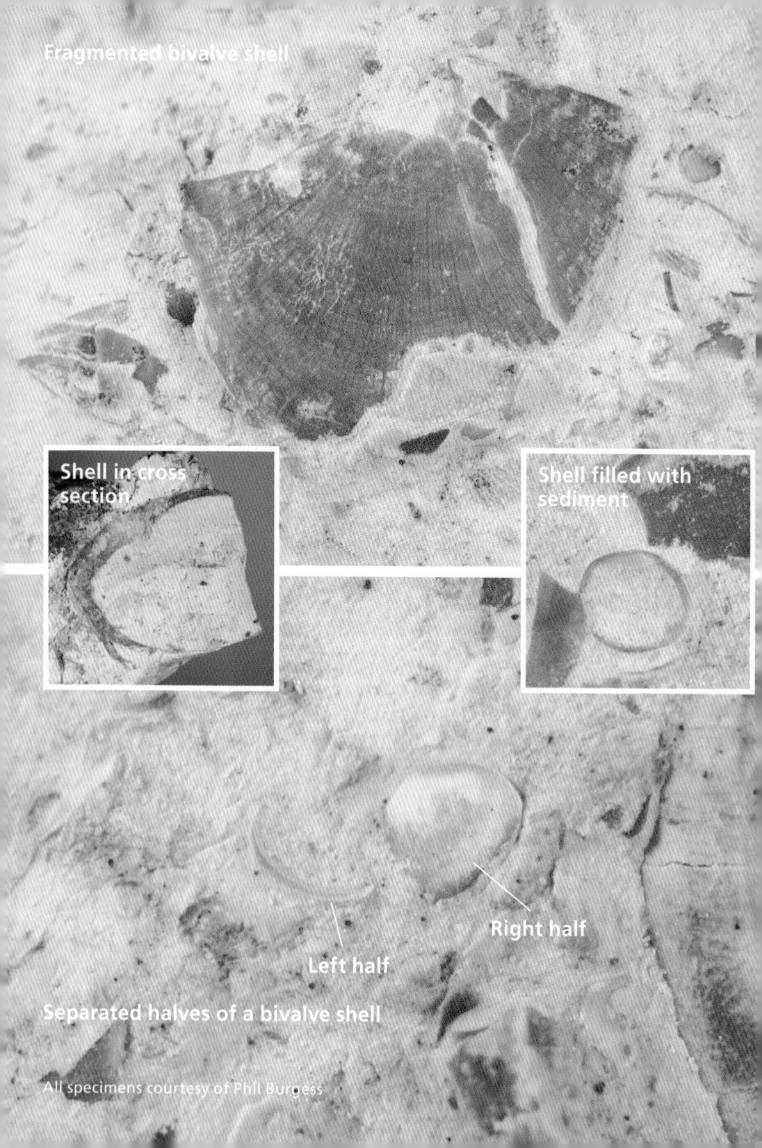

All specimens courtesy of Phil Burgess

Fossils, Bivalves

HARDNESS: N/A **STREAK:** N/A

Primary Occurrence

ENVIRONMENT: Fields, rivers, outcrops, road cuts, quarries

WHAT TO LOOK FOR: Rocks, particularly limestone, containing structures similar to modern-day clam shells

SIZE: Bivalve fossils are typically smaller than your palm

COLOR: Tan to brown or gray; colored like the surrounding rock

OCCURRENCE: Uncommon

NOTES: Of all the many kinds of fossils of marine animals that can be found in the three states, bivalves are often one of the most easily identified because of their striking similarity to living examples. Bivalves are mollusks whose bodies are protected by two shells connected by a hinge; familiar modern-day examples include clams, mussels and oysters. They are an extremely long-lived class of animals that has lived in Earth's waters for over half a billion years, and many different species are found embedded in the area's sedimentary rocks as fossils. Most bivalve shells are dome-like in shape and are typically smooth in texture, with fine ridges along their surfaces. Bivalves orient themselves vertically, and the left and right shell are identical, which is a key trait that distinguishes them from brachiopods (page 111), which orient themselves horizontally and have two differently shaped shell halves. However, in many cases both halves of a bivalve shell are not present together, so learn to spot their typically smooth, rounded shells. Brachiopods tend to exhibit deeper ridges on their shells' surfaces.

WHERE TO LOOK: Eastern Iowa and southwestern Wisconsin are rich with fossils of all kinds. Near the Mississippi River, any rock outcrop and river gravel may yield specimens where the water has cut through sedimentary rock layers.

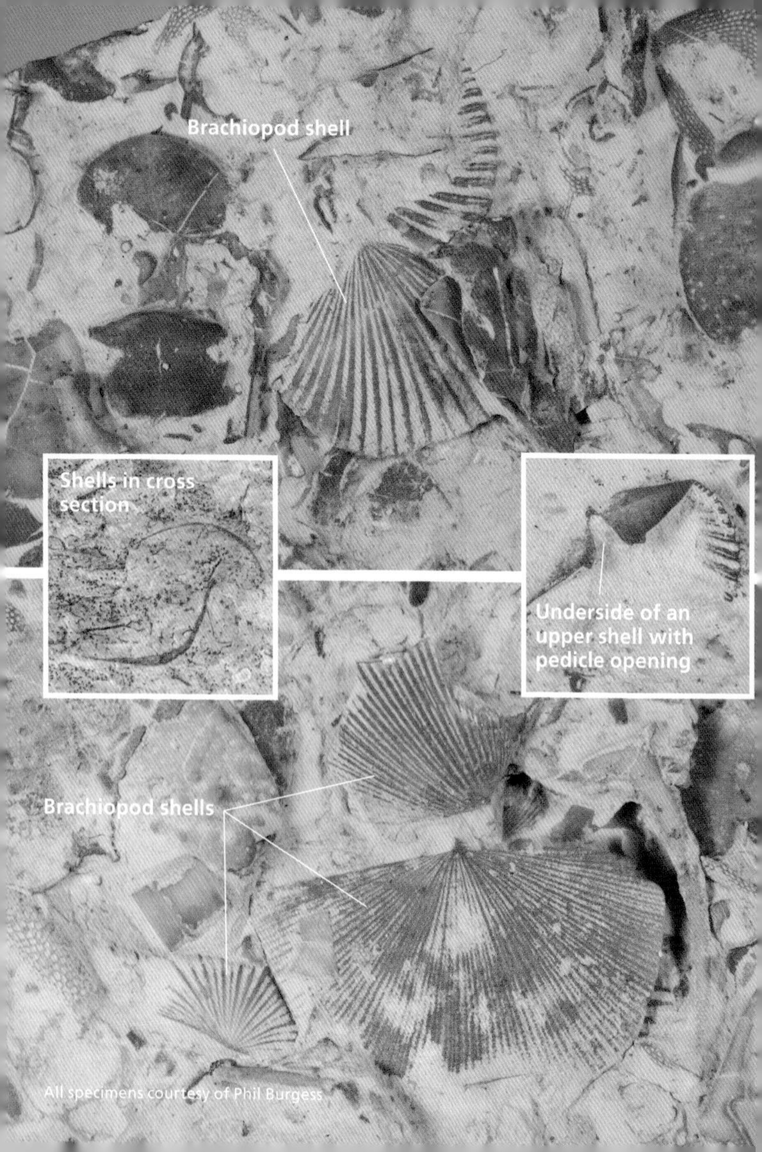

All specimens courtesy of Phil Burgess

Fossils, Brachiopods

HARDNESS: N/A **STREAK:** N/A

Primary Occurrence

ENVIRONMENT: Fields, rivers, outcrops, road cuts, quarries

WHAT TO LOOK FOR: Rocks, particularly limestone, containing structures similar to modern-day clam shells

SIZE: Brachiopod fossils are typically smaller than your palm

COLOR: Tan to brown or gray; colored like the surrounding rock

OCCURRENCE: Uncommon

NOTES: Brachiopods may initially appear to be bivalves (page 109), like clams, but they are actually quite different animals. While they do have two hard shells connected by a hinge that enclose their bodies, they have a special opening at the rear where a long muscle called a pedicle extends. The pedicle is used by some species as a "foot" and in others as an anchor. They are also different from bivalves in that they orient themselves horizontally and have upper and lower shells that are not identical. Often, but not always, the upper half is dome-shaped while the lower half is flatter. This is in contrast with bivalves' identical left- and right-oriented shells. While still alive today in our oceans, ancient brachiopod fossils can be found in this region's limestone, and their shells can be easily identified. Even if the asymmetrical halves of a brachiopod are not present or visible, you can distinguish them from bivalves by noting their often much more prominent grooves and ridges on their surfaces and by their less rounded and more fan-shaped structure.

WHERE TO LOOK: Eastern Iowa and southwestern Wisconsin are rich with fossils of all kinds. Near the Mississippi River, any rock outcrop and river gravel may yield specimens where the water has cut through sedimentary rock layers.

Straight-shelled cephalopod fossil
Specimen courtesy of Phil Burgess
Straight-shelled cephalopod fossil
Shell segments
Specimen courtesy of Phil Burgess

Fossils, Cephalopods

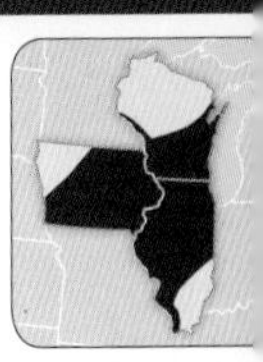

Primary Occurrence

HARDNESS: N/A **STREAK:** N/A

ENVIRONMENT: Fields, outcrops, road cuts, quarries

WHAT TO LOOK FOR: Rocks containing straight or coiled, segmented shell-like structures

SIZE: Cephalopod fossils can be quite large; they are often palm-sized and rarely even several feet in length

COLOR: Tan to brown or gray; colored like the surrounding rock

OCCURRENCE: Rare

NOTES: Modern-day octopuses and squid belong to an interesting class of mollusks called the cephalopods, and many species of their ancient ancestors had hard protective shells. Cephalopod shells are therefore some of the many fossils found in the region. Examples of cephalopod fossils include the coiled, almost snail-like shells of ammonites and the long, straight shells of other cephalopods. Like other shell fossils, they were originally composed of aragonite (page 49). But when they died and their shells joined the rest of the sediment on the seafloor, pressure and chemical changes over time recrystallized the aragonite into calcite (page 65). This is the same process by which limestone (page 155) forms, and it's therefore the primary host rock for cephalopod fossils. Straight-shelled cephalopod fossils are perhaps the most common in the region and typically not difficult to identify because their long, tapered shells can be conspicuous, unless fragmented. These shells are more-or-less oval shaped in cross section, and virtually always exhibit clearly delineated sections derived from the periodic growth of the animal. In crude finds, these details may be obscured, but their coarse, tube-like shapes are generally still apparent.

WHERE TO LOOK: Eastern Iowa and southwestern Wisconsin are rich with fossils. Near the Mississippi River, any rock outcrop or river gravel may yield specimens.

Various species of snail fossils

Specimens courtesy of Phil Burgess

Small snail shell embedded in chert

Specimen courtesy of Phil Burgess

Fossils, Gastropods

HARDNESS: N/A **STREAK:** N/A

Primary Occurrence

ENVIRONMENT: Fields, rivers, outcrops, road cuts, quarries

WHAT TO LOOK FOR: Limestone or chert containing spiral-shaped, snail shell-like structures

SIZE: Gastropod fossils are typically smaller than your fist

COLOR: Tan to brown or gray; colored like the surrounding rock

OCCURRENCE: Uncommon

NOTES: Gastropods are a class of mollusk that includes modern-day snails. One of the longest-lived groups of animals, they have been present on Earth for nearly half a billion years. As one of the many types of aquatic fossils that can be found in the region, gastropod shells are often one of the most rewarding discoveries, because they are directly comparable to a living animal and are therefore generally easy to identify. Found embedded in sedimentary rocks, especially limestone (page 155), but also chert and sandstone, gastropod shells can vary greatly in size, from smaller than ⅛ inch to the size of your fist, depending on the species. The shape of the shell also depends on the species; many exhibit a tightly spiraled habit while others are loosely coiled. A gastropod shell could possibly be confused with that of an ammonite, a variety of cephalopod (page 113) with a coiled-shell, but ammonites are much rarer, and they are also often more complex in structure, exhibiting ridges between shell segments. Most shells made by ancient snails are smoother in texture and not as clearly segmented.

WHERE TO LOOK: Eastern Iowa and southwestern Wisconsin are rich with fossils of all kinds. Near the Mississippi River, any rock outcrop and river gravel may yield specimens where the water has cut through sedimentary rock layers.

Water-worn petrified wood
Specimen courtesy of Phil Burgess

Fossils, Plants

HARDNESS: 3–4 **STREAK:** Greenish white

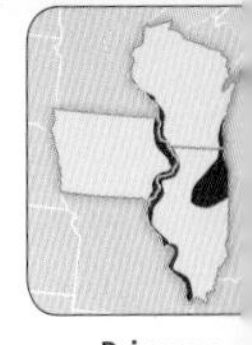

Primary Occurrence

ENVIRONMENT: Fields, rivers, outcrops, road cuts, quarries, lakeshores

WHAT TO LOOK FOR: Rocks containing the indications of plants, such as leaf-like shapes and wood grain

SIZE: Plant fossils can be up to palm-sized and rarely larger

COLOR: Tan to brown or gray; colored like the surrounding rock

OCCURRENCE: Uncommon

NOTES: When organic materials fossilize, it is often the hard parts of animals that are preserved—bones, shells, teeth, etc. But the soft and delicate parts of plants may be mineralized as well, sometimes retaining amazing amounts of detail. Petrified wood, popular with collectors, is an example of tree matter that has been fossilized; it was typically replaced by quartz-based material, such as chert (page 79). This means that a lot of petrified wood is very hard and weather resistant, enabling specimens to survive being transported great distances by rivers and glaciers. In addition, fine details like wood grain and bark may also be preserved. Concretionary nodules are another type of plant fossil found in the region; they consist of hard, round rock formations that contain a fossil at their core. Concretions (page 89) formed around organic matter when the decaying material reacted with the surrounding rock, cementing it together around the fossil.

WHERE TO LOOK: Petrified wood can be widespread, as the glaciers brought fragments from far and wide, even Canada; look in river gravel, especially along the Mississippi. The finest fossil plant nodules originate from shale in the Mazon Creek area near Morris in northeastern Illinois. Concretions can be found in the banks along the river, but many areas are privately owned and some are protected.

Stromatolites in limestone
Arching stromatolite structures
Stromatolite cast
Specimen courtesy of Phil Burgess
Stromatolite "waves" in jasper
Large stromatolite layers in chert
Specimen courtesy of Phil Burgess

Fossils, Stromatolites

HARDNESS: N/A **STREAK:** N/A

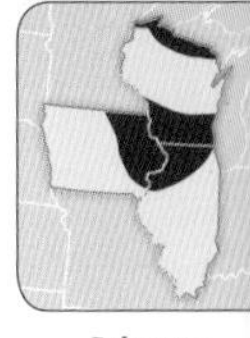

Primary Occurrence

ENVIRONMENT: Lakeshores, rivers, quarries, mines

WHAT TO LOOK FOR: Rocks exhibiting layered, curving features embedded within, often with a mushroom-like shape

SIZE: Stromatolite formations are embedded in rock and can be quite large, up to basketball-sized or even larger

COLOR: Varies greatly; often multicolored in shades of tan to brown, red, gray to black, and less commonly green

OCCURRENCE: Uncommon

NOTES: Over the eons, inland seas repeatedly submerged these three states; those seas teemed with many forms of ancient life, traces of which can still be found today. Some of the oldest you can find are stromatolites, which are the preserved remains of cyanobacteria colonies. Still in existence today, this blue-green algae lives in "blobs" that rise upward from seafloors; their sticky surfaces accumulate rock and mineral sediments as they grow, producing a towered, sometimes mushroom-shaped structure. As sediments settled around ancient stromatolites and turned to rock, the algae's mineral-rich layers were preserved. Now, their characteristic curved, sweeping layered shapes can be seen within various sedimentary rocks, including limestone (page 155), chert (page 79), jasper (page 151), and even in banded iron formation (page 57). Quartz-rich stromatolites, especially those in jasper, are frequently confused with agates (page 37), but are always more opaque.

WHERE TO LOOK: Stromatolites in jasper can be found in glacial till all over the region, but primarily in Wisconsin, near Lake Superior, and in banded iron formations near Hurley. Stromatolites can also be seen in limestone from eastern Iowa, southwestern Wisconsin and all over northern Illinois.

Burrows in limestone

Burrows in limestone

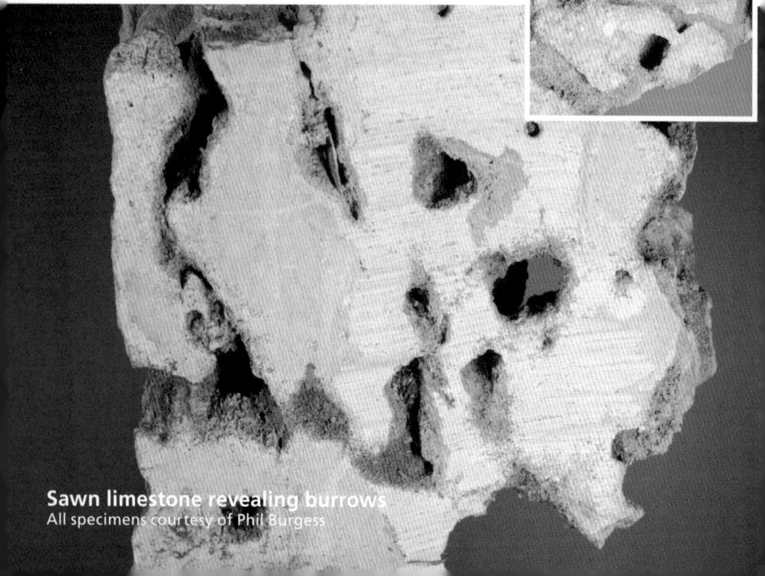

Sawn limestone revealing burrows

All specimens courtesy of Phil Burgess

Fossils, Trace

HARDNESS: N/A **STREAK:** N/A

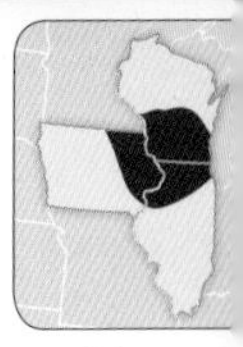

Primary Occurrence

ENVIRONMENT: Fields, outcrops, road cuts, quarries

WHAT TO LOOK FOR: Limestone with unusual tunnel-like holes

SIZE: Trace fossils can be nearly any size, up to several feet

COLOR: Tan to brown or gray; colored like the surrounding rock

OCCURRENCE: Uncommon

NOTES: When we think of fossils, we tend to imagine pieces of ancient life—bones, teeth and leaves. But some fossils are more subtle, exhibiting not what is there, but what was there. These are called trace fossils, and they are preserved evidence of an ancient animal's behavior, rather than the animal itself. Trace fossils include footprints, burrows and even feces. Though several kinds of trace fossils are present in the region, only burrows and bore holes are abundant. These tunnel-like holes in rock, particularly limestone (page 155), were made by aquatic animals, particularly sea cucumbers, as they burrowed into seafloor sediments. As the sediments then solidified into rock, the burrows were preserved. Somewhat ironically, in many cases the animal that made the holes had no hard body parts and therefore itself was not preserved at all. Differentiating between a burrow and simple hole in the rock can of course be difficult; look for long, meandering pathways that don't seem to have been made by erosion. Research and expert help will then be your next step toward positive identification.

WHERE TO LOOK: The sedimentary rock formations known as the Galena Group and the Platteville Group, located in northern Illinois, southwestern Wisconsin, and northeastern Iowa, are riddled with bore holes and burrows created by ancient sea life. Other trace fossils, such as animal tracks, have been found in central Wisconsin.

Trilobite tail segment

Trilobite head segment

Loose spine

Tail segment

Trilobite tail segment

All specimens courtesy of Phil Burgess

Fossils, Trilobites

HARDNESS: N/A **STREAK:** N/A

Primary Occurrence

ENVIRONMENT: Fields, rivers, outcrops, road cuts, quarries

WHAT TO LOOK FOR: Limestone containing traces of segmented, almost insect-like animals with bulbous eyes

SIZE: Trilobite fossils are typically smaller than your palm

COLOR: Tan to brown or gray; colored like the surrounding rock

OCCURRENCE: Rare

NOTES: There are seemingly countless types of ancient aquatic life to be found as fossils in the region's sedimentary rocks, but few are quite as exciting as the trilobites. Though present in Earth's oceans for almost 300 million years, this class of arthropod was unable to survive a mass-extinction event at the end of the geological period known as the Permian. Today, trilobites' closest living relatives are crustaceans and arachnids. Trilobites had hard-shelled bodies composed of several armor-like segments and a head bearing two bulbous compound eye structures, each containing numerous tiny lenses. As trilobites evolved, their bodies became more complex, developing features such as spikes. As fossils, trilobites are conspicuous when found whole; their oval-shaped bodies consist of segments that look like repeating ridges and are almost washboard-like. But disjointed pieces of their shells are more abundant, and identification of these often requires an expert. Their heads (called the cephalon), however, are more apparent, often exhibiting a dome- or horseshoe-like shape with two bumps—their compound eyes.

WHERE TO LOOK: Eastern Iowa and southwestern Wisconsin are rich with fossils of all kinds. Near the Mississippi River, any rock outcrop and river gravel may yield specimens where the water has cut through sedimentary rock layers.

Fulgurite fragments
Glassy interior
Glassy interior
Hollow, glassy interior
Rough, sandy exterior

Fulgurites

HARDNESS: <6.5 **STREAK:** N/A

RANDOM

Primary Occurrence

ENVIRONMENT: Lakeshores, rivers, outcrops

WHAT TO LOOK FOR: Hard, ragged, glassy tube-like structures found in sand or gravel, often with branching shapes

SIZE: Fulgurites can range greatly in size, but specimens longer than an inch or two are very rare

COLOR: Commonly gray to tan or brown; more rarely greenish

OCCURRENCE: Very rare

NOTES: When lightning strikes the earth, its immense energy is transferred into the ground where its superheated plasma travels downward, melting and fusing the material it passes through. When that material is sand, the affected grains melt and resolidify in an instant, cooling to form a mass of natural glass called lechatelierite. Lechatelierite formed by a lightning strike is called a fulgurite, nicknamed "petrified lightning." Lechatelierite lacks a true crystal structure (a trait shared by all glasses) and therefore is not actually a mineral, but because sand consists mostly of quartz, lechatelierite is in turn composed mostly of silica (quartz material), which makes fulgurites quite hard and brittle. They take the form of elongated, twig-like structures, often with branching arms, a reflection of the lightning that created them. Their exteriors are typically rough and coated in gritty sand or gravel that stuck to the glass while it was molten. Their interiors are usually hollow and glassy. But finding a fulgurite can be challenging, to say the least; finding one is often a matter of luck.

WHERE TO LOOK: Because of the random nature of lightning strikes, it is impossible to say where you may find a fulgurite. Hilltops or sandy areas with tall trees may be your best bet; try looking in the dirt around burned trees or shrubs.

Galena crystals

Galena

HARDNESS: 2.5 **STREAK:** Lead-gray

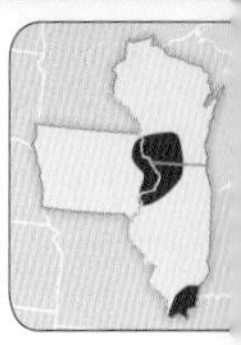

ENVIRONMENT: Mines, quarries

WHAT TO LOOK FOR: Very soft, dark, metallic material that is very heavy; often exhibits a markedly blocky structure

SIZE: Most pieces of galena are smaller than your palm

COLOR: Dark lead-gray, sometimes with a white to tan coating

OCCURRENCE: Uncommon in WI and IL; rare in Iowa

NOTES: Galena is the primary ore of lead and played a major role in the mining histories of Illinois and southern Wisconsin. It consists of a simple combination of lead and sulfur and formed within the region's limestone as the rock underwent chemical changes. Usually darkly colored and metallic, it is brightly lustrous when broken or freshly exposed, but many weathered specimens are coated in a dull white or tan crust of other lead-bearing minerals, such as cerussite (page 71). Often found within cavities in limestone alongside calcite or fluorite, crystals take the form of perfect cubes, or blocky clusters and masses with step-like features. Veins are also common and may not exhibit crystal structure, but even these specimens are easy to identify as galena is very soft and has a high specific gravity, which means that even small specimens feel heavy for their size. In addition, all specimens, even crude ones, have cubic cleavage, and will break into perfect cubes when carefully struck. Despite its lead content, galena is safe to handle, but don't inhale its dust.

WHERE TO LOOK: The Hardin County, Illinois, mines are legendary for specimens, but the mines are mostly off-limits. Limestone cavities in Grant and Lafayette counties, as well as around Mineral Point, are productive areas for galena in Wisconsin. Galena is rare in Iowa but can be found in limestone in many counties bordering the Mississippi River.

Garnet in schist
Garnets in granite
Garnets in quartz
Clusters of tiny garnets in granite

Garnet Group

HARDNESS: 6.5–7.5 **STREAK:** Colorless

Primary Occurrence

ENVIRONMENT: Outcrops, road cuts, rivers, quarries, lakeshores, mines

WHAT TO LOOK FOR: Very hard, small, dark ball-like crystals embedded in granite or schist

SIZE: Garnets are typically smaller than a pea

COLOR: Dark red to brown

OCCURRENCE: Uncommon in Wisconsin; rare in IA and IL

NOTES: Consisting of several minerals of a similar composition and structure, the garnet group is a large family of minerals that form in a number of ways, typically as a constituent of granite (page 143) or as a product of metamorphism in gneisses and schists (page 135). They are typically reddish or brown in color and develop as small ball-like crystals with many facets (faces), which are often diamond-shaped. In this region, only Wisconsin is home to garnets in their host rock; garnets found in the state include andradite, spessartine, grossular, and, the most abundant of them, almandine. Trying to tell them apart is often fruitless for amateurs, so simply identifying them as "garnet" is generally sufficient, and this can be accomplished by looking for their trademark shape, high hardness, reddish coloration, and association with certain rocks. Garnets are also very weather-resistant, and when weathering frees them from their host rock they often end up in sand or gravel.

WHERE TO LOOK: Northeastern Wisconsin has ample opportunities to collect garnets, such as within slate in Florence County, and further south in Marathon County, where they can be found embedded in granite. Elsewhere in the region, any river may yield pebbles of granite or other rocks bearing garnets, and tiny ones may be found loose in sand or gravel.

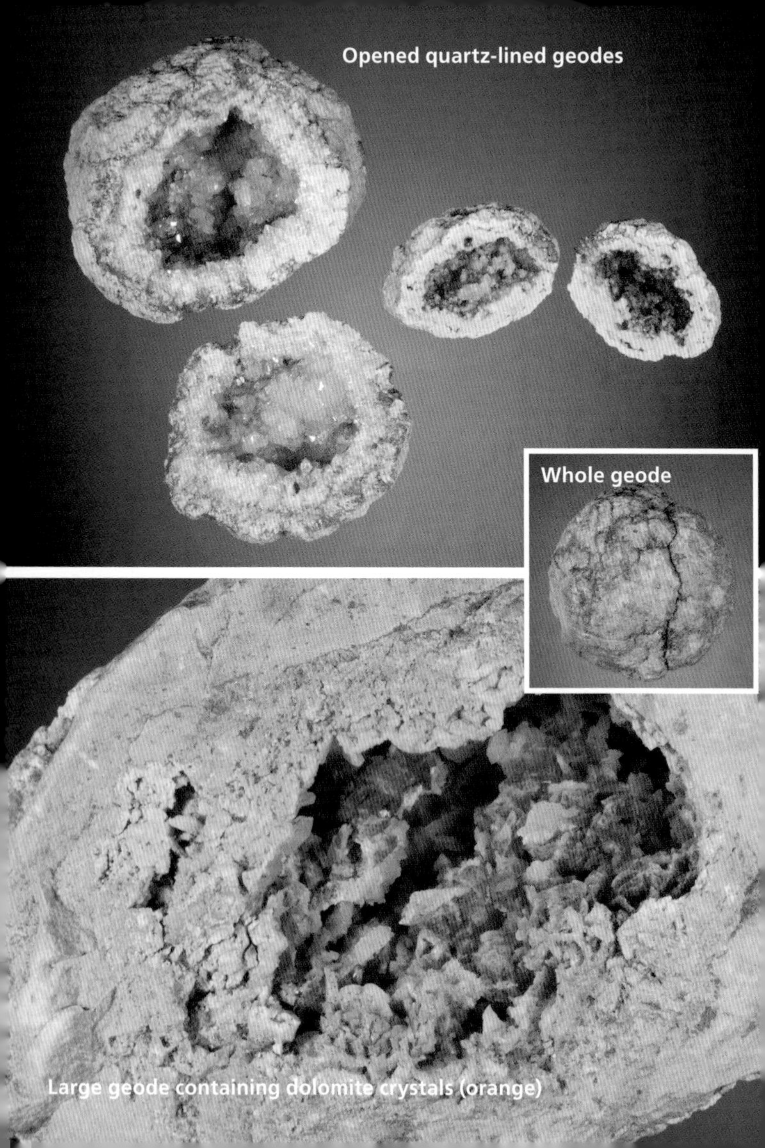
Opened quartz-lined geodes
Whole geode
Large geode containing dolomite crystals (orange)

Geodes

HARDNESS: N/A **STREAK:** N/A

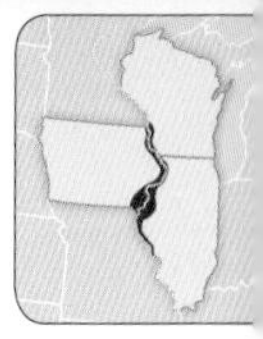

Primary Occurrence

ENVIRONMENT: Quarries, mines, outcrops, fields, rivers

WHAT TO LOOK FOR: Spherical rocky masses that are hollow when broken open; often with interiors lined by crystals

SIZE: Most geodes are fist-sized or smaller, but they can rarely be up to a foot or more in size

COLOR: Varies greatly; exteriors are often brown to tan or gray

OCCURRENCE: Uncommon

NOTES: If there is one collectible for which the upper Midwest is best known, it's geodes. A geode is a rounded rock formation with a hollow center and interior walls typically lined with mineral crystals, namely quartz and calcite. There are several types of geodes found around the world, but in this region the term applies primarily to those that formed in limestone within cavities created by anhydrite or fossils (there is some dispute about which). Often nearly spherical in shape, these tan balls of rock are fairly conspicuous, whether they are still embedded in their host rock or when weathering has freed them. But their appeal isn't obvious until they are carefully cut open or broken apart; a variety of crystals are found within them. Quartz and calcite are common finds, but dolomite, bitumen, millerite, clay minerals, marcasite and a number of other minerals may be found in the void, all generally finely crystallized and highly desirable.

WHERE TO LOOK: The most famous region for geodes is around Keokuk, Iowa, and Hamilton, Illinois, where geodes are common and a number of pay-to-dig quarries operate for collectors. Nearby rivers may also produce minor amounts of geodes weathered out of their host rock along the banks.

Glauconite grains (green) in sandstone
Algae on sandstone
for comparison
Layered "greensand"

Glauconite

HARDNESS: 2 **STREAK:** Colorless

Primary Occurrence

ENVIRONMENT: Rivers, quarries, fields, outcrops

WHAT TO LOOK FOR: Tiny green grains within sandstone

SIZE: Individual glauconite grains are just fractions of an inch

COLOR: Green to blue-green

OCCURRENCE: Uncommon in Wisconsin; rare in IA and IL

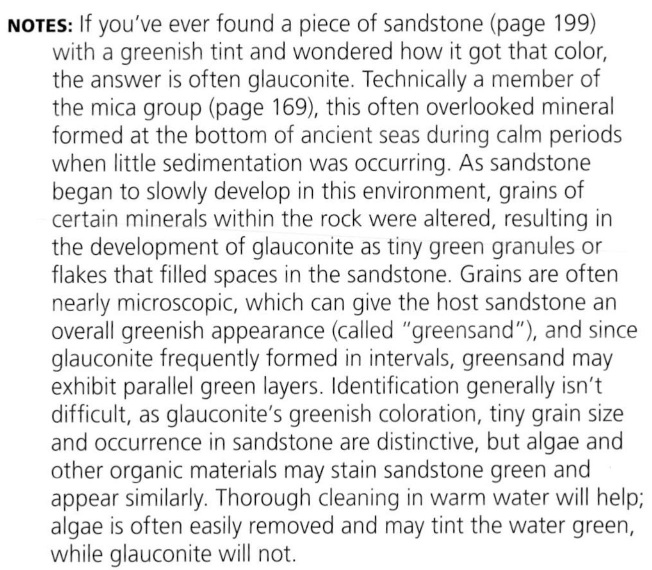

NOTES: If you've ever found a piece of sandstone (page 199) with a greenish tint and wondered how it got that color, the answer is often glauconite. Technically a member of the mica group (page 169), this often overlooked mineral formed at the bottom of ancient seas during calm periods when little sedimentation was occurring. As sandstone began to slowly develop in this environment, grains of certain minerals within the rock were altered, resulting in the development of glauconite as tiny green granules or flakes that filled spaces in the sandstone. Grains are often nearly microscopic, which can give the host sandstone an overall greenish appearance (called "greensand"), and since glauconite frequently formed in intervals, greensand may exhibit parallel green layers. Identification generally isn't difficult, as glauconite's greenish coloration, tiny grain size and occurrence in sandstone are distinctive, but algae and other organic materials may stain sandstone green and appear similarly. Thorough cleaning in warm water will help; algae is often easily removed and may tint the water green, while glauconite will not.

WHERE TO LOOK: In southwestern Wisconsin, where the Mississippi River cuts through sandstone, glauconite-rich layers are apparent in Pierce, St. Croix, and Vernon counties, and particularly near Pepin. Glauconite is rarer in Iowa and Illinois but can be found near their Wisconsin borders.

Granitic gneiss
Coarse layering

Schists
Tight layers
Mica schist

Gneiss/Schist

HARDNESS: N/A **STREAK:** N/A

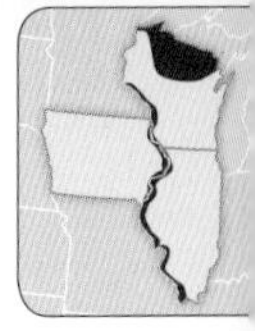

Primary Occurrence

ENVIRONMENT: Outcrops, road cuts, mines, quarries

WHAT TO LOOK FOR: Hard, dense, layered rocks of varying grain size that often contain pockets of very hard minerals

SIZE: As rocks, gneiss and schist can be found in any size

COLOR: Varies; typically multicolored white to gray, black, green

OCCURRENCE: Common in Wisconsin; uncommon in IA and IL

NOTES: When rocks are subjected to heat and/or pressure, they undergo physical and chemical alterations called metamorphism. This often partially melts the rocks, causing significant changes to their structure and mineral make-up; after this, the rocks then resolidify. Gneiss (pronounced "nice") is a perfect example of this process and is a general term used to describe rocks with loosely defined layers of minerals that were rearranged and compressed. Often described as having less than half of its minerals concentrated into layers, gneiss retains much of the original rock's appearance. Varieties are named for their parent rock, so granitic gneiss, for example, formed from granite and therefore retains many of its traits (but with noticeable layering). Schists, on the other hand, were so sufficiently metamorphosed that the original rock was completely changed, resulting in fairly hard, tightly compacted layers of new minerals. Schists are defined by their predominant mineral; mica schists are among the most abundant and consist primarily of "glittery" mica minerals, with possible clusters of hard minerals, like garnets.

WHERE TO LOOK: As with most other metamorphic rocks, gneiss and schist generally won't be found in Iowa or Illinois except as pebbles transported there by glaciers. But the northern half of Wisconsin has ample amounts of each, especially in western Marathon County and southern Ashland County.

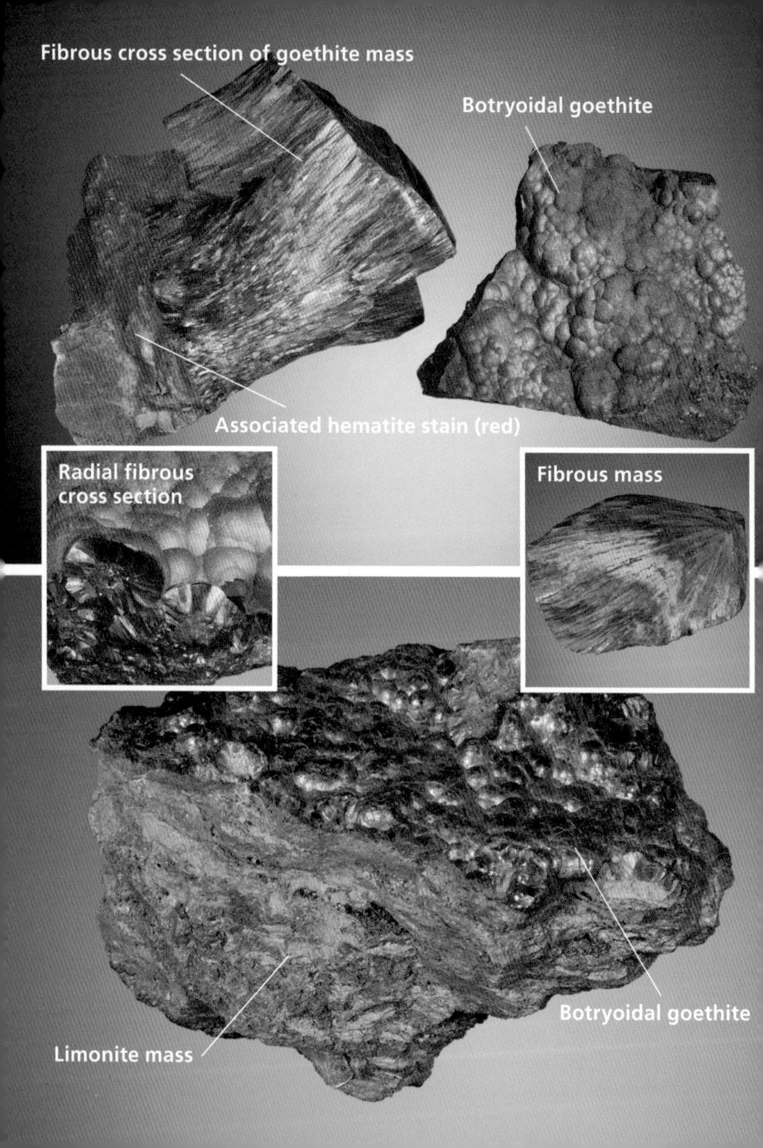
Fibrous cross section of goethite mass
Botryoidal goethite
Associated hematite stain (red)
Radial fibrous cross section
Fibrous mass
Botryoidal goethite
Limonite mass

Goethite

HARDNESS: 5–5.5 **STREAK:** Yellow-brown

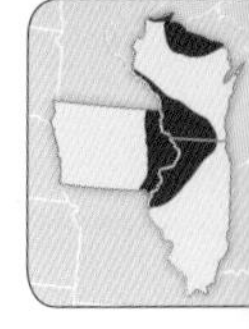

Primary Occurrence

ENVIRONMENT: All environments

WHAT TO LOOK FOR: Metallic brown mineral crusts with a yellowish surface coating or small, black needle-like crystals

SIZE: Masses can be very large, up to many feet in size; crystals are rarely larger than an inch

COLOR: Metallic black to brown; yellow-brown to rusty orange-brown when weathered

OCCURRENCE: Common in WI and IA; uncommon in IL

NOTES: Goethite (pronounced "gur-tite") is a common water-bearing ore of iron and formed in many different ways, but primarily as a result of other iron-bearing minerals breaking down. Metallic and typically dark brown to black, goethite often weathers to develop a yellow or rusty orange surface coating, which is often the first clue to its presence. It takes many forms in this region, including botryoidal (lumpy, grape-like) crusts, radiating fibers and stalactites in iron deposits. It also occurs as tiny, pointed needle-like crystals in geodes or limestone cavities, but it is also the primary constituent of ever-present limonite (page 157) and is therefore most abundant as earthy, granular crusts and stains on rocks and in soil. Massive specimens often have a fibrous cross section. Since hematite (page 147) shares many of these traits, confusing them is easy, but hematite has a reddish streak. Magnetite (page 159) is also similar but is magnetic.

WHERE TO LOOK: Goethite is fairly prevalent within cavities in limestone throughout the region. Eastern Iowa quarries and outcrops frequently contain black needle-like crystals, as do northwestern Illinois areas bordering the Mississippi River. In Wisconsin, the Gogebic Range of Ashland and Iron counties has produced copious amounts from mines and outcrops.

Goethite pseudomorphs after pyrite octahedrons

Specimen courtesy of Phil Burgess

Goethite pseudomorphs after marcasite crystal aggregates

Specimen courtesy of Phil Burgess

Goethite, Pseudomorphs

Primary Occurrence

HARDNESS: 5–5.5 **STREAK:** Yellow-brown

ENVIRONMENT: Fields, quarries, outcrops, mines

WHAT TO LOOK FOR: Metallic brown crystals shaped like pyrite or marcasite crystals

SIZE: Crystals can be found in masses up to several inches

COLOR: Metallic black to brown; yellow-brown to rusty orange-brown when weathered

OCCURRENCE: Uncommon

NOTES: Goethite's dark brown metallic crystals and botryoidal (lumpy, grape-like) masses may be a common sight in Wisconsin's iron-mining regions, but huge portions of the region are predominantly overlain by sedimentary rocks and they too have their own unique forms of goethite. Goethite pseudomorphs are one of these varieties. A pseudomorph is a mineral growth that forms when chemicals alter or replace a preexisting crystal. In our region, goethite pseudomorphs mimic marcasite (page 165) and pyrite (page 189) crystals. In the case of pyrite and marcasite, weathering causes these iron- and sulfur-bearing minerals to lose their sulfur content while their iron content changes forms to that found in goethite. However, throughout this process, the original crystal shapes of pyrite or marcasite are maintained. Therefore pseudomorphs are best identified by crystal shape. Pyrite has cubic or octahedral crystals while marcasite has curved points, often in serrated aggregates. After analyzing crystal shape, examine the specimen's color, streak and hardness.

WHERE TO LOOK: Quarries in the vicinity of the Mississippi River in southwestern Wisconsin and northeastern Iowa produce many interesting goethite pseudomorphs after marcasite, but beware that most collecting sites are privately owned. Prairie du Chien, Wisconsin, is one notable hunting locale.

Gold flakes freed from host rock (largest approx. 1/16")

Gold flake (approx. 1/64") on limonite

Specimen courtesy of Kevin Ponzio

Gold

HARDNESS: 2.5–3 **STREAK:** Metallic yellow

Primary Occurrence

ENVIRONMENT: Mines, rivers

WHAT TO LOOK FOR: Minute grains of lustrous yellow metal

SIZE: Most of the region's gold is found as tiny flecks smaller than 1⁄16 inch; most require magnification to see

COLOR: Always metallic yellow

OCCURRENCE: Very rare

NOTES: History's most iconic metal, gold is a valuable mineral pursued in virtually every state, including those not particularly known to yield it. Such is the case in Wisconsin, Iowa and Illinois, where minute amounts of the yellow metal can rarely be recovered. As a native element, gold is found uncombined with any other elements and therefore always exhibits a unique set of traits that make it easily distinguishable from any other mineral. It is deposited in a number of environments from various processes, but when mined from rock, it is typically found on or in quartz, and other materials, such as limonite. Wishful thinking often leads collectors to misidentify similarly colored metallic minerals such as pyrite (page 189), chalcopyrite (page 77) and marcasite (page 165), but all of these are much harder and extremely brittle. Gold, on the other hand, is malleable. Pressing the tip of a pin into a piece of gold will yield a small dimple or hole; pyrite, chalcopyrite and marcasite won't show the same result and will instead break and splinter. Finally, gold may be found in many rivers as tiny nuggets or flakes, called placer (pronounced "plasser") gold. When freed from their host rock, these grains will still be as easy to identify.

WHERE TO LOOK: The region is not widely known for opportunities to find gold, so there are no "sure bet" localities. Rivers in central Wisconsin may be your first place to try looking.

Granite
Diorite
Texture detail
Various kinds of granite
Feldspar
Mica

Granite

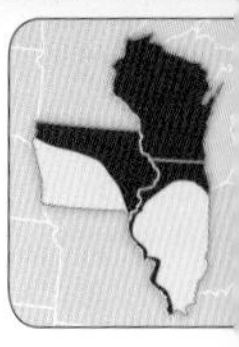

Primary Occurrence

HARDNESS: N/A **STREAK:** N/A

ENVIRONMENT: Quarries, road cuts, outcrops, lakeshores, rivers, fields

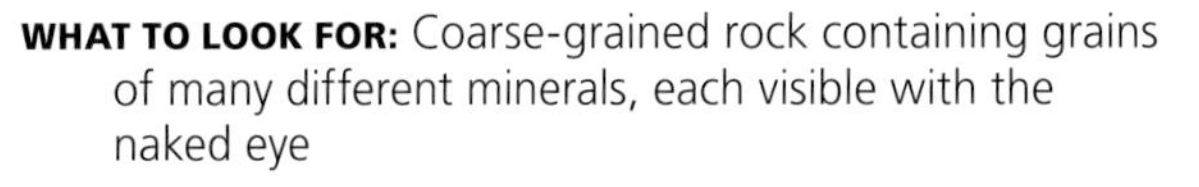

WHAT TO LOOK FOR: Coarse-grained rock containing grains of many different minerals, each visible with the naked eye

SIZE: As a rock, granite can be found in any size

COLOR: Varies greatly; multicolored, primarily in shades of tan, orange or red to pink, white to gray, brown, and black

OCCURRENCE: Very common

NOTES: A perfect example of an igneous rock, granite formed when magma (molted rock) buried deep within the earth cooled very slowly. Because granite "cooked" for a long time, its minerals crystallized to a large, visible size, giving granite its characteristic mottled appearance. In contrast, Rhyolite (page 201) contains largely the same minerals—primarily quartz, orthoclase feldspar, amphiboles and micas—but cooled very quickly on the earth's surface, resulting in a very fine-grained texture. Granite's mineral grains are fairly uniform in size, and it is predominantly light colored with darker minerals present as small spots. But when the ratio of granite's minerals changes, it is classified as a different type of rock. Diorite is an example of a granite-like rock that has more dark minerals than true granite. In this region, granite and diorite are found almost anywhere thanks to the glaciers, which carried boulders southward from Canada and northern Wisconsin.

WHERE TO LOOK: Granite is prevalent in glacial gravel all over the region, and boulders can be found far from their origin. But granite bedrock is only found in Wisconsin, particularly in the north-central part of the state, where exposures are common; the Wausau area has many granite quarries.

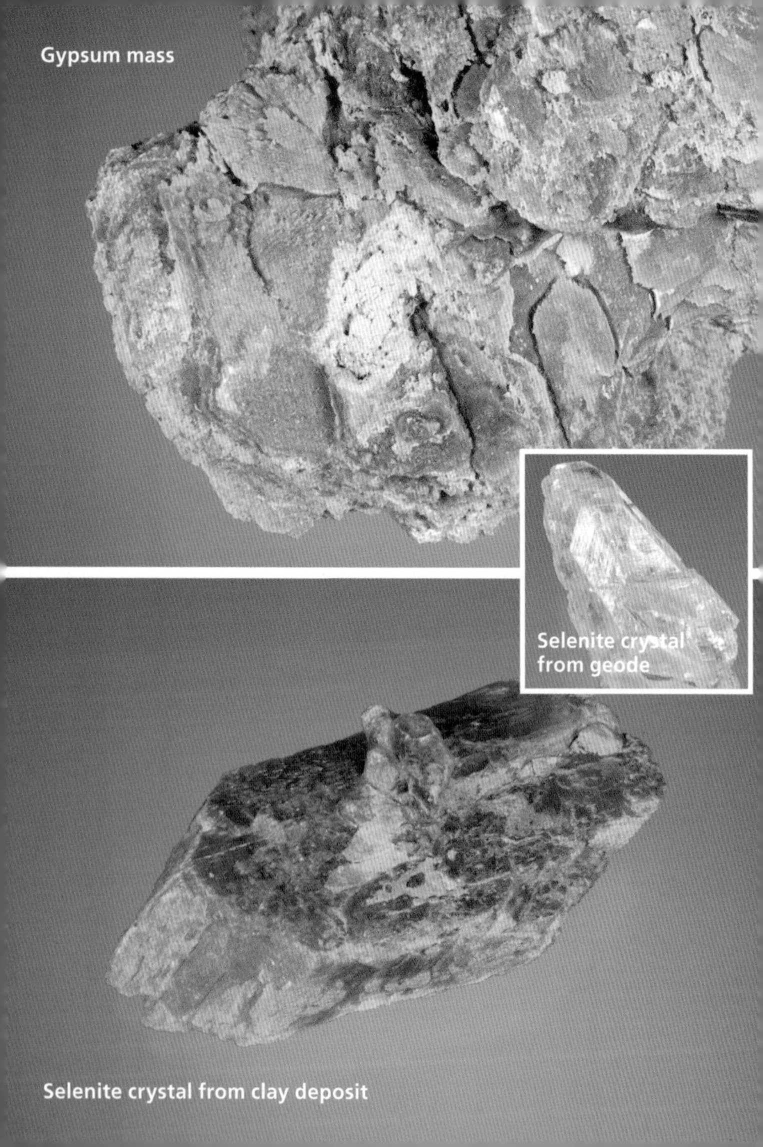

Gypsum mass

Selenite crystal from geode

Selenite crystal from clay deposit

Gypsum

HARDNESS: 2 **STREAK:** White

Primary Occurrence

ENVIRONMENT: Mines, outcrops

WHAT TO LOOK FOR: Very soft, light-colored masses or glassy angular crystals that are easily scratched by a fingernail

SIZE: Crystals may be up to palm-sized; masses may be up to several feet in size

COLOR: Colorless to white or gray, often stained yellow to orange or brown

OCCURRENCE: Common

NOTES: Used for centuries as the primary ingredient in plaster, gypsum is one of the most abundant sulfur-bearing minerals on earth and can form in a number of different ways. In our area, gypsum developed when ancient seas evaporated and left minerals behind. Large beds of massive gypsum are present in the region and are found between layers of sedimentary rocks such as shale. Such massive deposits of gypsum are typically opaque, chalky and granular in texture. Selenite, the name given to crystallized, translucent gypsum, is far more collectible, and it can be found as glassy sheets, blocky angular crystals, fibrous masses or delicate needles; selenite often formed when marcasite (page 165) and other sulfur-bearing minerals weathered, which freed their sulfur content. Whatever its form, gypsum is generally very easy to identify thanks to its very low hardness—you can scratch it with a fingernail—and its gradual solubility in water. It may resemble massive calcite (page 65) or aragonite (page 49), but both are harder than gypsum.

WHERE TO LOOK: Marion, Madison and Dallas counties in Iowa have produced well-formed crystals, often from coal mining areas. Most of Illinois' finest selenite specimens originated from the now-closed Hardin County mining district.

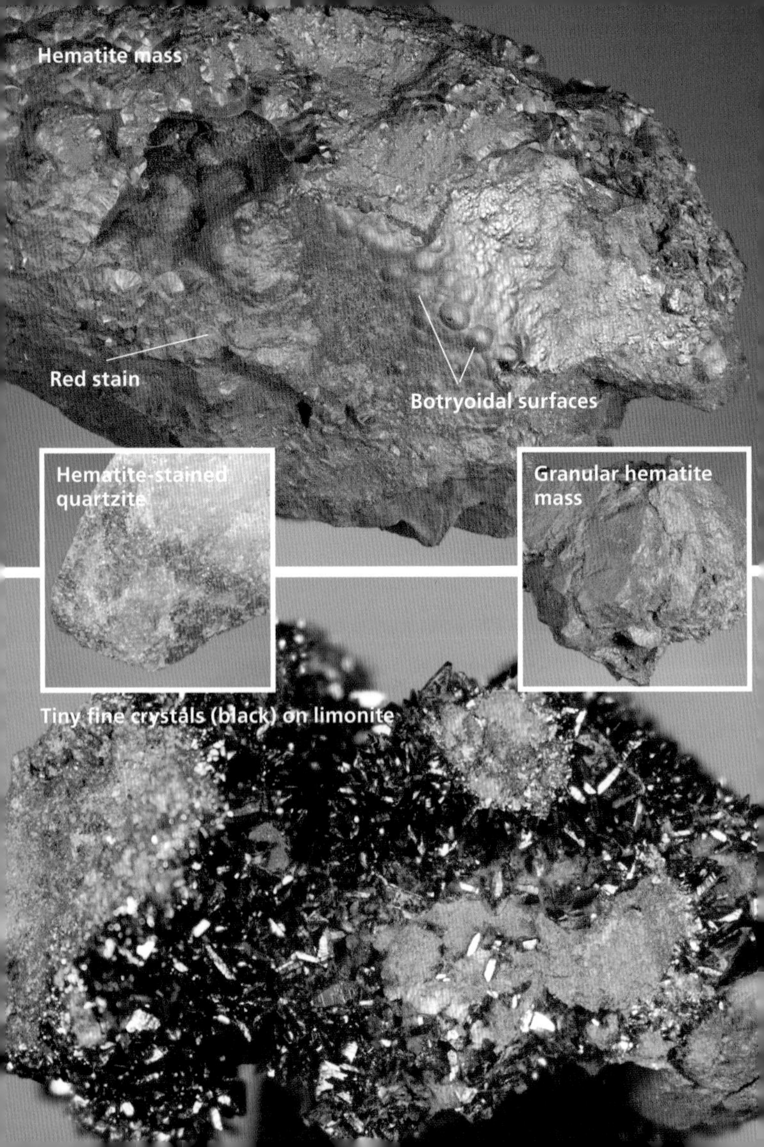
Hematite mass
Red stain
Botryoidal surfaces
Hematite-stained quartzite
Granular hematite mass
Tiny fine crystals (black) on limonite

Hematite

HARDNESS: 5–6 **STREAK:** Reddish brown

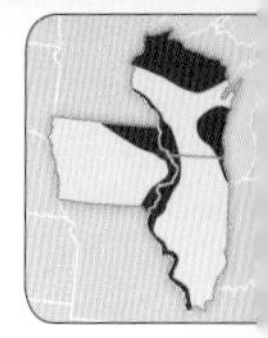

Primary Occurrence

ENVIRONMENT: All environments

WHAT TO LOOK FOR: Dark gray metallic mineral, often as lumpy crusts with a reddish-brown surface coating

SIZE: Masses of hematite can vary greatly in size; specimens are typically fist-sized or smaller but can grow to several feet

COLOR: Steel-gray to black; often stained red or reddish brown

OCCURRENCE: Common

NOTES: As a simple combination of iron and oxygen and the most abundant ore of iron on earth, hematite can form in a wide range of environments and takes many different forms. Crystals are quite rare but take the form of thin, hexagonal (six-sided) plates or dagger-like points; botryoidal (lumpy, grape-like) masses with fibrous interiors are more common. But the most common hematite finds of all are irregular veins or granular masses filling spaces in various rocks, or metallic crusts on rock surfaces. In these cases, hematite is virtually always opaque and dark metallic gray to black, very often with a rusty reddish brown surface coating. It is easy to confuse visually with magnetite (page 159), but hematite is not magnetic. Goethite also looks alike (page 137), but hematite has a reddish brown streak as opposed to goethite's yellowish brown streak. And since hematite turns reddish when weathered or when formed as microscopic grains, this color is a clue to its presence, even within other minerals; in red jasper (page 151), for example, hematite is the colorant.

WHERE TO LOOK: Hematite-bearing rocks may be found in glacial till anywhere throughout the region, especially along rivers. The region's finest specimens have come from Iron County, especially the famous Montreal Mine, where botryoidal, crystalline and granular hematite have been found.

Ilmenite grains (metallic) in gabbro
Ilmenite grains from heavy river sand

Ilmenite

HARDNESS: 5–6 **STREAK:** Brownish black

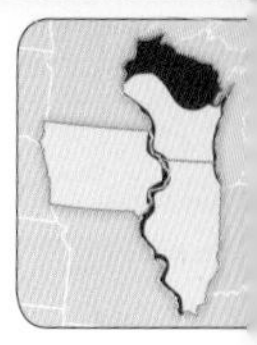

Primary Occurrence

ENVIRONMENT: All environments

WHAT TO LOOK FOR: Brittle, weakly magnetic, blue-black metallic grains embedded in dark rocks or loose in sand

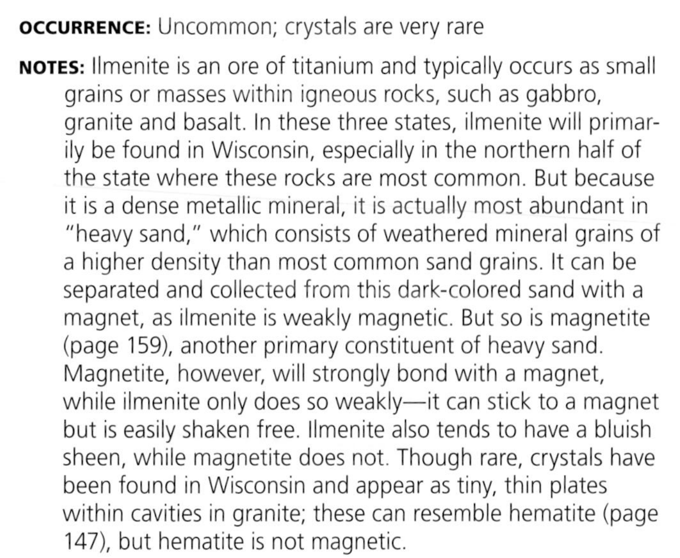

SIZE: Most ilmenite specimens are smaller than ⅛ inch

COLOR: Metallic black to bluish black

OCCURRENCE: Uncommon; crystals are very rare

NOTES: Ilmenite is an ore of titanium and typically occurs as small grains or masses within igneous rocks, such as gabbro, granite and basalt. In these three states, ilmenite will primarily be found in Wisconsin, especially in the northern half of the state where these rocks are most common. But because it is a dense metallic mineral, it is actually most abundant in "heavy sand," which consists of weathered mineral grains of a higher density than most common sand grains. It can be separated and collected from this dark-colored sand with a magnet, as ilmenite is weakly magnetic. But so is magnetite (page 159), another primary constituent of heavy sand. Magnetite, however, will strongly bond with a magnet, while ilmenite only does so weakly—it can stick to a magnet but is easily shaken free. Ilmenite also tends to have a bluish sheen, while magnetite does not. Though rare, crystals have been found in Wisconsin and appear as tiny, thin plates within cavities in granite; these can resemble hematite (page 147), but hematite is not magnetic.

WHERE TO LOOK: Most ilmenite will be found in river or beach sand, especially in northern Wisconsin and all along the Mississippi River, particularly around Fayville in southern Illinois. Gabbro in Ashland County, Wisconsin, contains large amounts of embedded ilmenite, and rare, small crystals have turned up in granite quarries near Rib Mountain in Wausau.

Iron-stained jasper
Jasper vein in rhyolite
Jasper vein in granite

Banded iron formation
with red jasper layers
Jasper with wave-like stromatolite structures

Jasper

HARDNESS: ~7 **STREAK:** White

Primary Occurrence

ENVIRONMENT: Lakeshores, rivers, fields, quarries

WHAT TO LOOK FOR: Very hard, opaque masses of reddish material, often with a smooth, waxy feel and appearance

SIZE: Masses of jasper can be found in a wide range of sizes, but most specimens will be smaller than your fist

COLOR: Varies greatly; red to brown is most common but can be yellowish to green or multicolored

OCCURRENCE: Common

NOTES: Jasper has long been a popular collectible because it is abundant, colorful and easily identified. Though they often form via different means, jasper is, in the simplest terms, the more colorful variety of chert (page 79). As such, it is a variety of microcrystalline quartz (page 195), consisting of microscopic granular quartz grains. This means that jasper has no distinct structure of its own and instead takes on a shape dictated by its surroundings, primarily irregular masses, nodules (round masses) or veins. Like any form of quartz, jasper is extremely hard and exhibits conchoidal fracturing (when struck, circular cracks appear), and it can withstand a great deal of weathering. When worn it has smooth, waxy textured surfaces and luster; freshly broken pieces are rougher and duller in appearance. Jasper is always more opaque than chalcedony (page 73), which is translucent but with which it can be confused. Much of the region's jasper is found in rivers and on lakeshores as small smooth pebbles, which were carried south by glaciers.

WHERE TO LOOK: Any area that saw glaciation may yield easily collected pebbles of jasper, especially in northern Wisconsin, near Lake Superior. The Mississippi and other major rivers also yield countless specimens in their gravel beds.

Slag
Specimen courtesy of Phil Burgess

Junk

HARDNESS: N/A **STREAK:** N/A

Primary Occurrence

ENVIRONMENT: All environments

WHAT TO LOOK FOR: Unnatural materials that may mimic rocks or minerals

SIZE: Junk can be found in any size

COLOR: Varies greatly; may be any color

OCCURRENCE: Very common

NOTES: An often overlooked element of rock and mineral collecting is the tendency to find man-made materials (such as cement, scrap metal, slag and other junk) that may initially appear to be naturally occurring materials. A weathered, water-worn piece of cement, for example, may at first glance resemble conglomerate (page 91), though cement is generally much harder. Metallic materials, such as a mass of iron, may be rusty, pitted and magnetic, leading many wishful thinkers to believe they have found a meteorite. Other metals may appear much stranger; melted soda cans show up on shorelines as shiny aluminum "blobs." But perhaps the biggest culprit for fooling collectors is slag, a man-made glassy substance formed as a by-product of industry. Often peppered with bubbles, slag may appear to be a volcanic rock, or, to optimists, some kind of rare mineral. In reality, no mineral will exhibit bubbles like those of slag.

WHERE TO LOOK: Invariably, wherever people have ventured, junk will follow. Man-made materials, including aluminum blobs, often collect on lakeshores, particularly along Lake Superior, as well as along any river; the Mississippi is, unfortunately, quite polluted in some areas. Any current or former quarries and mine sites will also be full of junk, though some things, such as old mining artifacts, may actually be somewhat collectible.

Limestone fragments
Texture detail
Water-worn pebble
Snail shell from limestone
Limestone
Dark dolostone

Limestone

HARDNESS: 3–4 **STREAK:** N/A

Primary Occurrence

ENVIRONMENT: All environments

WHAT TO LOOK FOR: Soft but tough, compact light-colored rock with a chalky feel that is abundant in flatter regions

SIZE: As a rock, limestone can be found in any size

COLOR: Commonly white to tan or brown, also light to dark gray

OCCURRENCE: Very common

NOTES: As the most common rock in the region, you won't have any trouble finding limestone. It formed at the bottom of the shallow marine seas that inundated the area millions of years ago, forming from the remains of microscopic organisms, animal shells and coral that settled into thick beds. These organic sediments, which consisted of unstable aragonite (page 49), slowly converted to calcite (page 65) and solidified into limestone—which consists of over 50 percent calcite, along with dolomite, clay and a small amount of quartz. Due to its high calcite content, limestone is fairly soft and easily scratched with a knife, and it will effervesce (fizz) in acids as weak as undiluted vinegar. Typically white, tan or gray, it develops a chalky white color and texture when a specimen has been worn, particularly on weathered edges. All of these traits will help you distinguish it from other sedimentary rocks. Limestone is also a common host to fossils of ancient sea life and often has cavities filled with fine crystals of calcite and other minerals. Finally, a dolomite-rich variety of limestone called dolostone (or "dolomite-rock") is also prevalent in the region but is generally visually identical.

WHERE TO LOOK: Limestone is extremely prevalent, including in the whole of Iowa and Illinois and the southern half of Wisconsin. The Mississippi River crosscuts enormous formations.

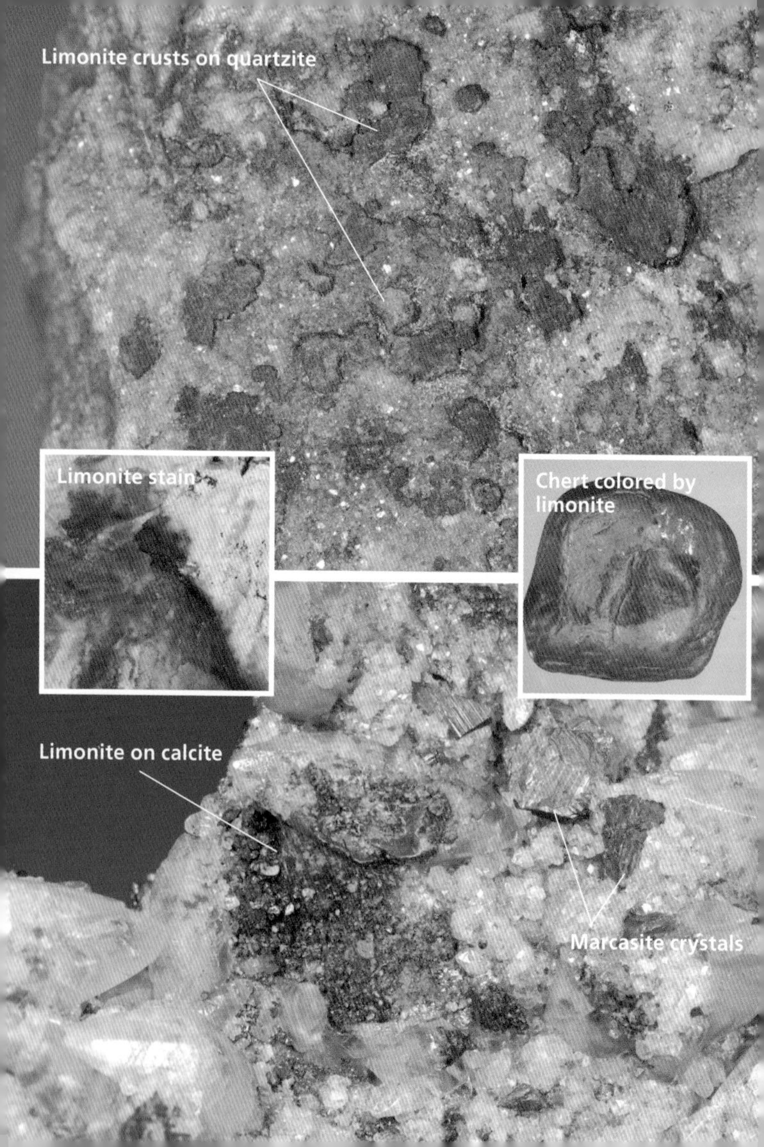
Limonite crusts on quartzite
Limonite stain
Chert colored by limonite
Limonite on calcite
Marcasite crystals

Limonite

HARDNESS: ~4–5.5 **STREAK:** Yellowish brown

Primary Occurrence

ENVIRONMENT: Quarries, outcrops, road cuts, fields, mines

WHAT TO LOOK FOR: Chalky, grainy yellow-brown crusts on rock, or irregular masses with no discernible crystal structure

SIZE: Masses of limonite can measure up to several inches in size

COLOR: Yellow to brown and orange to rust colors are common; rarely metallic brown to black

OCCURRENCE: Very common

NOTES: "Limonite" is not technically a mineral; instead, it is a general term used to describe granular mixtures of various water-bearing iron minerals, especially goethite (page 137). It forms as a result of iron-bearing minerals weathering and eroding, and is therefore most often found as a chalky, dusty coating or stain on the surface of other minerals, particularly sulfur-bearing iron minerals like pyrite (page 189). It can also be found in almost any type of rock, and it is especially common as a stain on porous rocks like sandstone. Masses large enough to collect are uncommon. Because limonite has no crystal structure, it can appear similar to goethite and have an identical streak, but limonite is generally softer and won't have crystals or a fibrous cross section like goethite will. Finally, limonite is so prevalent that it can be assumed present in any rock, sand or soil with an orange-brown coloration or stain, including yellow jasper (page 151).

WHERE TO LOOK: Limonite can be found almost anywhere rocks are exposed to weathering. Wisconsin's Gogebic Iron Range in the northeastern portion of the state, centered around the Hurley area, is rich with iron-bearing rocks and minerals and limonite is a common find.

Mass of magnetite
Octahedral crystal
Magnet attracted
to schist
Magnetite-rich schists

Magnetite

HARDNESS: 5.5–6.5 **STREAK:** Black

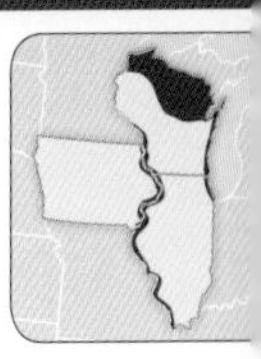

Primary Occurrence

ENVIRONMENT: All environments

WHAT TO LOOK FOR: Metallic black pyramid-shaped crystals, embedded masses or grains that are attracted to a magnet

SIZE: Specimens are generally smaller than ¼ inch

COLOR: Metallic black; rusty yellow to brown when weathered

OCCURRENCE: Common in Wisconsin; uncommon in IA and IL

NOTES: A primary mineral in most iron deposits, including banded iron formations (page 57), magnetite is an easy mineral to find and identify. It is always metallic black (unless weathered, in which case it may be more rust-colored), and well-formed crystals are octahedral in shape, resembling two four-sided pyramids placed base-to-base. But crystals are rare, and embedded grains or granular masses are far more common, in which case you'll have to check for its most diagnostic trait: magnetism. Not many minerals will attract a magnet, and magnetite is by far the most abundant of them. Magnetite is a particularly common constituent of dark rocks like gabbro and basalt, pebbles of which may be so magnetite-rich that the entire rock appears to be magnetic. When these rocks weather, the grains of magnetite are freed and concentrate in rivers and on beaches as black sand; a magnet will net you countless tiny specimens, including crystals. Visually, magnetite can be easily confused with hematite (page 147) and ilmenite (page 149), but hematite isn't magnetic and ilmenite is only weakly so.

WHERE TO LOOK: The Lake Superior region in northern Wisconsin is rich with magnetite. Iron-rich Ashland County will also yield many rocks attracted to a magnet. In Iowa and Illinois, finds are limited to grains in sand and gravel; dark patches of sand are common on Lake Michigan beaches near Chicago.

Specimen courtesy of Rob Carlson

Malachite on quartz druse
Specimen courtesy of Kevin Ponzio

Malachite

HARDNESS: 3.5–4 **STREAK:** Light green

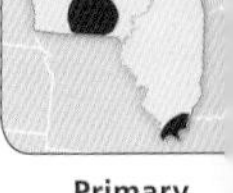
Primary Occurrence

ENVIRONMENT: Mines, quarries, outcrops

WHAT TO LOOK FOR: Soft, vivid green crusts or masses, sometimes with a fibrous structure, often alongside copper

SIZE: Malachite specimens are rarely larger than an inch or two

COLOR: Light to dark green

OCCURRENCE: Uncommon in Wisconsin and Illinois; rare in IA

NOTES: Malachite is a mineral that forms primarily when copper or copper-bearing minerals weather and decompose. You've probably already seen it many times; the green coating that forms on copper, including coins, is largely malachite. Though it can form in many environments and circumstances, its color is nearly always green, which is often a collector's first identifying clue. Crystals generally aren't present, especially in this region; they appear as tiny, delicate, slender prisms that are often needle-like in shape and arranged into small bundles or radial groupings. Irregular crusts, botryoidal (grape-like) masses, or poorly formed crystal clusters are much more common, especially near copper-bearing minerals, particularly chalcopyrite (page 77). These examples of malachite are often dusty and crumbly but may still exhibit a telltale fibrous appearance, especially on a freshly broken cross section. Poorly formed malachite may resemble chrysocolla (page 83), but chrysocolla is generally more blue and never forms crystals or is fibrous.

WHERE TO LOOK: Malachite is widespread in the region and is often found in cavities in sedimentary rocks resulting from weathered chalcopyrite. Waste rock at coal mines in Iowa has produced minor amounts, as have lead and zinc mines in Hardin County, Illinois. In Wisconsin, the copper-rich basalts of Douglas County also contain malachite.

Manganese oxide coating (black) on limestone

Specimen courtesy of Phil Burgess

Dendrites on limestone

Specimen courtesy of Phil Burgess

Manganese Oxides

HARDNESS: <6 **STREAK:** Black

Primary Occurrence

ENVIRONMENT: Fields, quarries, outcrops, mines, road cuts

WHAT TO LOOK FOR: Dusty black coatings or tree-like formations on the surfaces of rocks

SIZE: Coatings of manganese oxides may measure several inches

COLOR: Dark gray to black, occasionally brownish

OCCURRENCE: Common

NOTES: Just as limonite (page 157) consists of iron deposits accumulating on rocks, manganese-bearing compounds weather out of other minerals and make their way onto other materials. Typically found as just a thin dusty coating, often within a crack in limestone, "manganese oxides" are actually any one of a number of manganese- and oxygen-bearing minerals, such as psilomelane, pyrolusite and romanèchite (though distinguishing one from another is impossible outside of a lab). These black crusts or smudges of manganese minerals are normally rather mundane and easily ignored, but there is one highly collectible variety: dendrites. Identified by their branching, seemingly organic tree-like shapes, dendrites are microscopically thin and easily scratched away if not carefully collected. Occasionally misidentified as fossilized plants, dendrites are interesting curiosities, often quite small and not always easily noticed. Whatever their form, manganese oxides often leave a sooty black dust on your hands after handling, which is an easy way to detect their presence.

WHERE TO LOOK: Limestone formations in northern Illinois, southwestern Wisconsin and northeastern Iowa often hold manganese oxide coatings and dendrites in fissures and cracks in the rock. Some dendrites even formed on fossils.

Specimen courtesy of Kevin Ponzio

Marcasite

HARDNESS: 6–6.5 **STREAK:** Dark gray to black

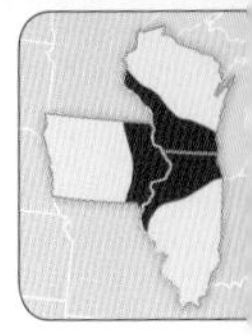
Primary Occurrence

ENVIRONMENT: Quarries, road cuts, outcrops, mines

WHAT TO LOOK FOR: A hard, brassy metallic mineral formed as plate-like crystals, often with deeply striated (grooved) faces

SIZE: Individual crystals are rarely larger than an inch, but crystal clusters or massive specimens may measure several inches

COLOR: Pale brass-yellow to grayish yellow common, also metallic brown, or with a reddish or multicolored surface coating

OCCURRENCE: Uncommon

NOTES: Marcasite consists of iron sulfide, a chemical compound made up of iron and sulfur. Pyrite (page 189) also consists of iron sulfide, but the two minerals differ because their crystal structures formed under varying conditions. And since their brassy colors are quite similar (though marcasite is often a bit duller or grayer in color), you'll have to recognize their shapes to tell them apart. While pyrite's typical shape is cubic or octahedral, marcasite forms elongated, tabular (broad) plates that are generally deeply striated (grooved) and frequently grow in clusters. Often, two or more marcasite crystals are twinned (grown within each other), and they form rectangular, hexagonal, or even serration-like shapes. You could also confuse it with chalcopyrite (page 77), but chalcopyrite is softer and forms triangular, pyramid-like crystals. Marcasite is most abundant in sedimentary rocks, particularly growing within cavities—often alongside calcite—in limestone and dolostone.

WHERE TO LOOK: Some of the best marcasite crystals in the country have come from southern Wisconsin, particularly in Grant and Lafayette counties. In Illinois, quarries along the Mississippi, such as Carroll County, have been lucrative, and in Iowa, marcasite is found in geodes from the Keokuk area.

Ball-like marcasite formations

Goethite pseudomorphs after marcasite crystal grouping
Specimen courtesy of Phil Burgess

Marcasite (continued)

HARDNESS: 6–6.5 **STREAK:** Dark gray to black

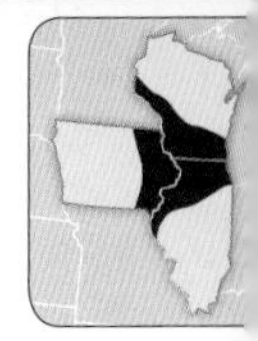
Primary Occurrence

ENVIRONMENT: Quarries, road cuts, outcrops, mines

WHAT TO LOOK FOR: A hard, brassy metallic mineral found in blade-like crystals or angular, ball-like masses

SIZE: Individual crystals are rarely larger than an inch, but crystal clusters or massive specimens may measure several inches

COLOR: Pale brass-yellow to grayish yellows are common; also found in metallic brown, or with a reddish or multicolored surface coating

OCCURRENCE: Uncommon

NOTES: Marcasite is fairly abundant in the sedimentary rocks of Iowa, southern Wisconsin, and Illinois, and it can take on a number of different forms, depending on the conditions present during its formation. In contrast to the beautiful bladed crystals typically sought after by collectors, marcasite can also form as crude ball-like formations with vague crystal points or as irregular masses embedded in rock, primarily limestone. As with any marcasite specimen, these may resemble pyrite (page 189), but marcasite tends to be grayer in color, often with a tinge of green. When collecting marcasite, keep this important note in mind: its crystal structure is unstable, which makes it prone to crumbling in collections. This is especially true for poorly formed specimens, and in high humidity environments marcasite will release sulfur as it degrades which may then affect nearby specimens. As an unstable mineral, marcasite is also frequently pseudomorphed into goethite (page 137); this is the result of an interesting chemical change that turns a specimen into goethite while retaining the crystal form of marcasite.

WHERE TO LOOK: The area around Prairie du Chien, Wisconsin, produces unique goethite pseudomorphs.

Crude mica crystal cluster from schist
Layered nature
Bright luster
Mica in granite
Mica schists

Mica Group

HARDNESS: 2.5–3 **STREAK:** Colorless

Primary Occurrence

ENVIRONMENT: All environments

WHAT TO LOOK FOR: Often dark-colored minerals that occur as thin, flexible sheets or tiny flakes that almost appear metallic

SIZE: Mica crystals are all paper-thin and typically very small, but masses in coarse-grained rocks may be an inch or two

COLOR: Gray to brown common, occasionally black or yellow

OCCURRENCE: Very common; crystals are only found in Wisconsin

NOTES: The mica group is a large family of minerals that all share a set of unique and easily identifiable traits. In these three states, only three micas are common enough to find: muscovite, phlogopite and biotite (a general name to describe otherwise unidentified dark colored micas). Each form as thin, highly flexible, hexagonal (six-sided), sheet-like crystals that grow in flaky, layered stacks called "books." And well-formed stacks are indeed book-like; in many examples, individual crystals are so flexible they can be peeled away like pages. Unfortunately, well-crystallized micas are rare, found primarily only in granite (page 143) or pegmatite (page 181), and most micas are instead found as tiny flecks in schist, shale, mudstone and other metamorphic or sedimentary rocks. In the latter examples, micas tend to give the rocks a "glittery" appearance, thanks to mica's most noticeable characteristic: its luster. Many specimens are so shiny that they may initially appear to be metallic; along with their low hardness and flaky nature, this makes them easy to identify.

WHERE TO LOOK: Micas are common as tiny reflective flakes in any shale, mudstone or schist throughout the region, and especially as larger dark flakes and crystals in granite. Larger crystals are only found in northern Wisconsin, particularly in the famous quarries in Marathon County, near Wausau.

Millerite cluster on calcite
Millerite in quartz
Hair-like crystals on calcite

Millerite

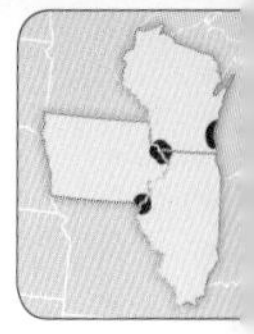

Primary Occurrence

HARDNESS: 3–3.5 **STREAK:** Greenish black

ENVIRONMENT: Quarries, mines, rivers, outcrops

WHAT TO LOOK FOR: Slender, hair- or needle-like brassy crystals, typically in divergent "spray"-shaped groupings

SIZE: Most millerite crystals are no longer than ½ inch

COLOR: Brass-yellow, sometimes with a slight greenish tinge

OCCURRENCE: Rare

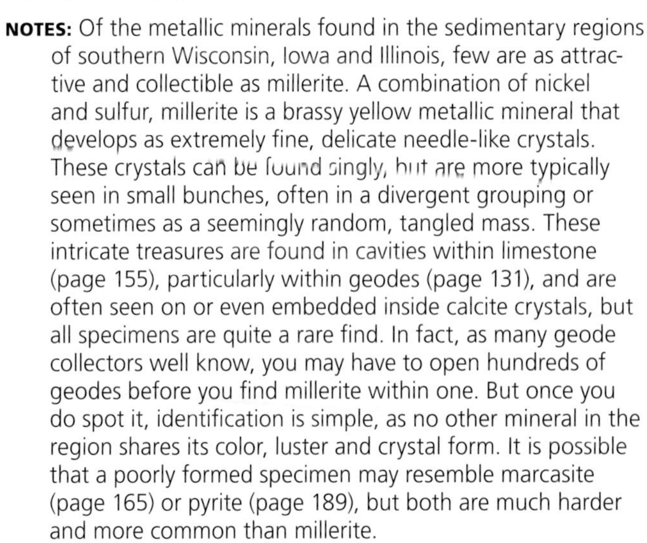

NOTES: Of the metallic minerals found in the sedimentary regions of southern Wisconsin, Iowa and Illinois, few are as attractive and collectible as millerite. A combination of nickel and sulfur, millerite is a brassy yellow metallic mineral that develops as extremely fine, delicate needle-like crystals. These crystals can be found singly, but are more typically seen in small bunches, often in a divergent grouping or sometimes as a seemingly random, tangled mass. These intricate treasures are found in cavities within limestone (page 155), particularly within geodes (page 131), and are often seen on or even embedded inside calcite crystals, but all specimens are quite a rare find. In fact, as many geode collectors well know, you may have to open hundreds of geodes before you find millerite within one. But once you do spot it, identification is simple, as no other mineral in the region shares its color, luster and crystal form. It is possible that a poorly formed specimen may resemble marcasite (page 165) or pyrite (page 189), but both are much harder and more common than millerite.

WHERE TO LOOK: The Hamilton, Illinois, and Keokuk, Iowa, area quarries are renowned for their geodes, which seldom contain fine millerite crystals. They remain your best bet for finding any specimen of this rare mineral.

Molybdenite (metallic) on granite
Close-up
Molybdenite (metallic) on granite

Molybdenite

HARDNESS: 1–1.5 **STREAK:** Grayish green; black on paper

Primary Occurrence

ENVIRONMENT: Quarries, mines, outcrops

WHAT TO LOOK FOR: Very soft, flaky, metallic gray veins or masses in granite; very rarely as thin, flexible six-sided crystals

SIZE: Masses can measure several inches, but are typically smaller than your palm; crystals are rarely wider than ½ inch

COLOR: Metallic gray, sometimes with a bluish tint

OCCURRENCE: Rare in Wisconsin; not found in Iowa or Illinois

NOTES: Molybdenite, consisting of molybdenum and sulfur, is one of the easiest metallic minerals in the region to identify. It most often found as thin veins or masses within granite (page 143) or its coarsely crystallized cousin, pegmatite (page 181). When found, molybdenite is conspicuous thanks to its metallic gray, sometimes bluish, color and brightly reflective luster. Masses develop in flaky, paper-thin layers often without a distinct shape; crystals are extremely scarce but are typically very small hexagonal (six-sided) plates, mostly embedded within quartz. Molybdenite's color can resemble galena's (page 127), but galena is harder, much heavier, cubic and generally found in limestone, not granite. Mica minerals (page 169) can also seem similar, due to their bright luster and similar crystal shape, but they aren't actually metallic. Micas are also far more common, translucent and don't occur as veins. But molybdenite is best identified by its extremely low hardness (you can scratch it with your fingernail) and its flexibility, especially in thin specimens.

WHERE TO LOOK: You won't find molybdenite in Iowa or Illinois, but it is collectible in Wisconsin. Granite outcrops in Marinette County have yielded large amounts of molybdenite specimens embedded in quartz or feldspar, and in Marathon and Florence Counties within pegmatites.

Moonstone fragment in bright light showing bright internal schiller

Blocky, cleaved masses of moonstone

Rectangular cleavage

"Moonstone"

HARDNESS: 6–6.5 **STREAK:** White

Primary Occurrence

ENVIRONMENT: Outcrops, quarries

WHAT TO LOOK FOR: Light-colored masses with internal parallel lines and a bluish schiller; masses break into rectangular pieces

SIZE: Masses of moonstone can be up to softball-sized or larger

COLOR: White to gray or tan, often with brown staining; blue to green or rarely gold-colored internal schiller (flashes)

OCCURRENCE: Rare in Wisconsin; not found in Iowa or Illinois

NOTES: Moonstone has been a highly desired collectible since the days of ancient Rome, but it isn't actually a distinct mineral. The term "moonstone" is used to describe various feldspar minerals (page 101) that exhibit a blue-to-green schiller, or "flashes" of light from within when rotated. This phenomenon is caused by the feldspar forming in thin, translucent, tightly bonded layers; when light enters these layers, it bounces between them, creating the schiller effect. Most moonstone from around the world is actually a combination of albite and orthoclase feldspars, but the moonstone for which the Wausau, Wisconsin, area is known consists of tightly layered anorthoclase, a rarer feldspar. Found as crude masses within granite, Wisconsin's moonstone exhibits all the traits of other feldspars: a high hardness, a rectangular, step-like cleavage, and feldspar's typically light coloration. Another identifying trait of moonstone is that freshly broken surfaces often reveal internal parallel lines resulting from its layered structure. But only feldspars with the bluish schiller are called moonstone.

WHERE TO LOOK: Moonstone is only found in Marathon County near Wausau in granite outcrops and quarries. The primary quarry where most specimens originated is closed to the public, and you'll need permission to visit.

Chert layers
Siltstone layers
Water-worn siltstone

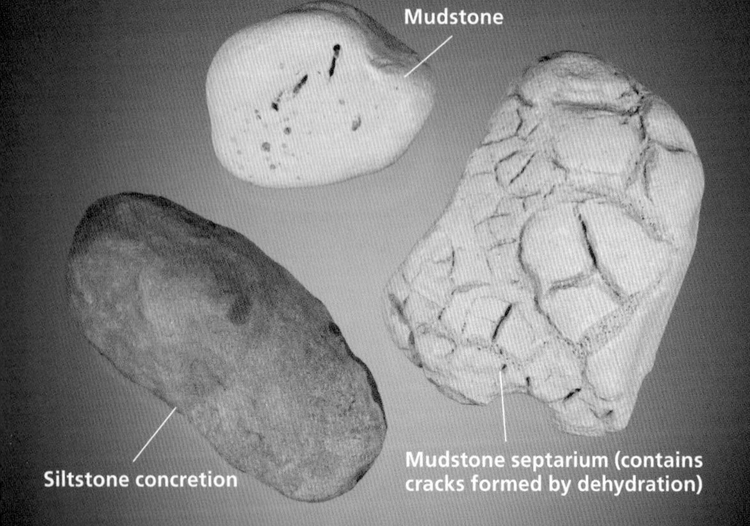
Mudstone
Siltstone concretion
Mudstone septarium (contains cracks formed by dehydration)

Mudstone/Siltstone

HARDNESS: N/A **STREAK:** N/A

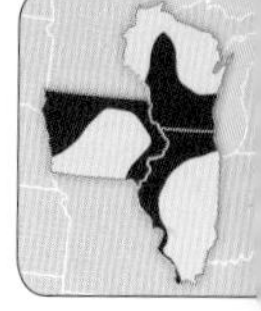

Primary Occurrence

ENVIRONMENT: Fields, quarries, rivers, road cuts, outcrops

WHAT TO LOOK FOR: Soft, dense rocks that resemble hardened clay and consist of nearly microscopic grains

SIZE: Both rocks can be found in any size

COLOR: Light to dark gray, tan to brown, sometimes reddish

OCCURRENCE: Common

NOTES: Sedimentary rocks are typically defined by the size of the weathered mineral particles they consist of; siltstone consists of tiny silt-sized particles, around 1/5,000 of an inch in size, while mudstone consists of microscopic mud-sized particles only 1/12,500 of an inch. When found, both rocks are extremely fine-grained, evenly colored and fairly soft, with a gritty feel. Both rocks also formed at the bottoms of ancient lakes or oceans. The difference between the two, other than the size of the sediments from which they are composed (which will only be observable under an extremely powerful microscope), is that mudstone contains large amounts of clay minerals (page 85). This may help in telling the two rocks apart, as mudstone tends to be a bit softer and may split apart more easily. Siltstone may also contain concretions (page 89) and layers or pockets of chert (page 79) embedded within it. Finally, shale (page 205) is actually largely the same as mudstone but is highly layered.

WHERE TO LOOK: Large portions of central and eastern Iowa are underlain by these rocks, as are many of the western Illinois counties that border the Mississippi River. In Wisconsin, Sawyer and Vilas counties contain large amounts of these rocks, much of which is partially metamorphosed to a harder, more compact state (called argillite).

Olivine grains from river sand (approx. 1/16")

Olivine Group

HARDNESS: 6.5–7 **STREAK:** Colorless

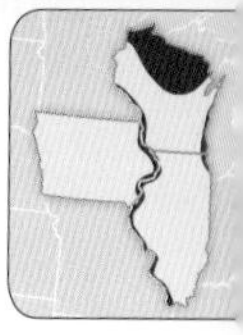

Primary Occurrence

ENVIRONMENT: Quarries, outcrops, road cuts, rivers, lakeshores, mines

WHAT TO LOOK FOR: Very hard, glassy green-to-yellow masses, often as small grains embedded in dark rocks

SIZE: Most specimens will measure smaller than ¼ inch

COLOR: Yellow to green or dark green; sometimes yellow-brown

OCCURRENCE: Common in Wisconsin; rare in IA and IL

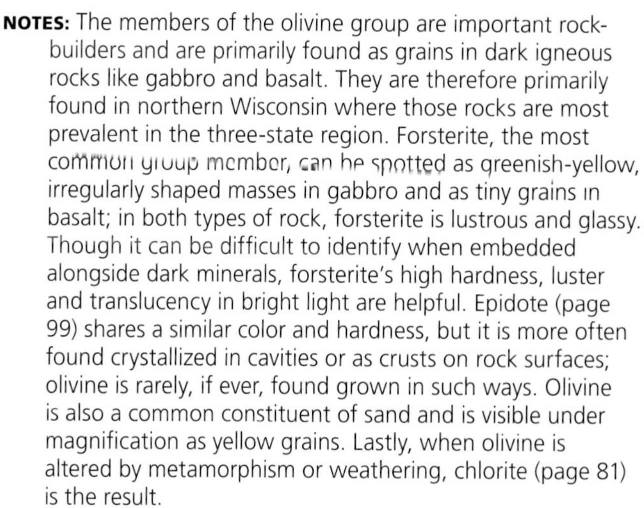

NOTES: The members of the olivine group are important rock-builders and are primarily found as grains in dark igneous rocks like gabbro and basalt. They are therefore primarily found in northern Wisconsin where those rocks are most prevalent in the three-state region. Forsterite, the most common group member, can be spotted as greenish-yellow, irregularly shaped masses in gabbro and as tiny grains in basalt; in both types of rock, forsterite is lustrous and glassy. Though it can be difficult to identify when embedded alongside dark minerals, forsterite's high hardness, luster and translucency in bright light are helpful. Epidote (page 99) shares a similar color and hardness, but it is more often found crystallized in cavities or as crusts on rock surfaces; olivine is rarely, if ever, found grown in such ways. Olivine is also a common constituent of sand and is visible under magnification as yellow grains. Lastly, when olivine is altered by metamorphism or weathering, chlorite (page 81) is the result.

WHERE TO LOOK: Although notable samples of olivine have been found in Iowa and Illinois meteorites, those occurrences are extremely rare. Olivines are really only found in northern Wisconsin—in gabbro in Ashland County, in basalt in Bayfield County and in glacial sand all over the region.

Large, well-formed microcline crystal from a pegmatite (approx. 6")

Pegmatite

HARDNESS: N/A **STREAK:** N/A

Primary Occurrence

ENVIRONMENT: Outcrops, quarries, road cuts

WHAT TO LOOK FOR: Extremely coarse-grained rock composed of large, well-formed crystals of various minerals

SIZE: As a rock, pegmatite can be found in any size

COLOR: Varies greatly; multicolored, typically in shades of white to tan, yellow to brown, pink, gray to black, or green

OCCURRENCE: Rare in Wisconsin; not found in IA or IL

NOTES: Pegmatite is a very coarse-grained rock that forms in close association with granite (page 143), deep within the earth where it is very hot. The high temperatures prevent magma (molten rock) from cooling quickly, allowing its minerals time to grow to a visible size. But pegmatite forms at the core of a granite formation or as dykes ("arms" of magma that fill fractures in pre-existing rocks) where the magma can be enriched with groundwater. The water helps mobilize mineral molecules, resulting in much larger mineral grains than those found in granite. Pegmatite typically contains the same minerals as granite—feldspars, micas, quartz, amphiboles—as well as other rarer minerals, like tourmaline (page 217). Of the three states, pegmatite is only found in Wisconsin, where scattered outcroppings represent the "roots" of the long-gone Penokean Mountain Range that once towered over the region. It can be identified by its granite-like coloration and the large, well-formed, angular mineral crystals embedded within it; being familiar with feldspars (page 101) will help.

WHERE TO LOOK: You'll only find pegmatite in Wisconsin, where it primarily occurs in the northeast portion of the state. Florence, Forest and Marinette counties contain numerous outcrops that yield collectible minerals, as do quarries in Marathon County, in the area around Wausau.

Phenakite crystal mass freed from granite (approx. ¼")

Tiny lustrous phenakite crystal on granite (approx. 1/16")

Phenakite

HARDNESS: 7.5–8 **STREAK:** White

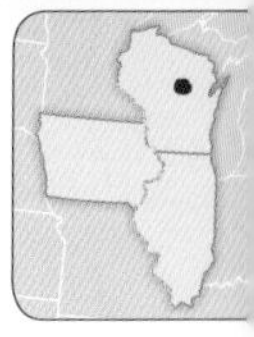
Primary Occurrence

ENVIRONMENT: Mines, quarries

WHAT TO LOOK FOR: Small, brightly lustrous, very hard and complexly shaped translucent white crystals

SIZE: Most phenakite crystals are smaller than 1/16 inch

COLOR: Colorless to white, sometimes yellowish

OCCURRENCE: Very rare in Wisconsin; not found in IA or IL

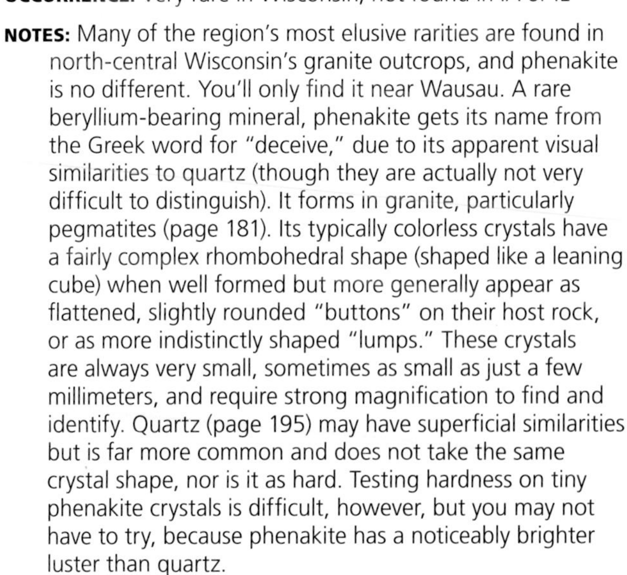

NOTES: Many of the region's most elusive rarities are found in north-central Wisconsin's granite outcrops, and phenakite is no different. You'll only find it near Wausau. A rare beryllium-bearing mineral, phenakite gets its name from the Greek word for "deceive," due to its apparent visual similarities to quartz (though they are actually not very difficult to distinguish). It forms in granite, particularly pegmatites (page 181). Its typically colorless crystals have a fairly complex rhombohedral shape (shaped like a leaning cube) when well formed but more generally appear as flattened, slightly rounded "buttons" on their host rock, or as more indistinctly shaped "lumps." These crystals are always very small, sometimes as small as just a few millimeters, and require strong magnification to find and identify. Quartz (page 195) may have superficial similarities but is far more common and does not take the same crystal shape, nor is it as hard. Testing hardness on tiny phenakite crystals is difficult, however, but you may not have to try, because phenakite has a noticeably brighter luster than quartz.

WHERE TO LOOK: Only granite quarries in Marathon County, Wisconsin, particularly near Wausau, produce these small, rare crystals; the area around Rib Mountain has been lucrative, but most of the productive quarries are privately owned and off-limits to collectors without permission.

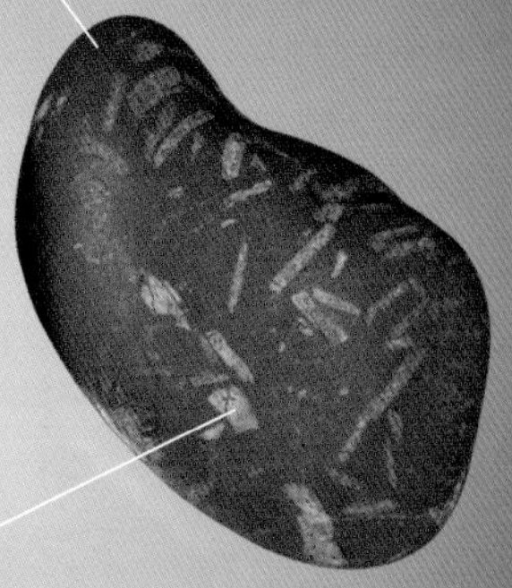
Basalt porphyry
Coarse rectangular crystals

Texture detail

Basalt porphyry with very large crystals

Rhyolite porphyry

Porphyry

HARDNESS: N/A **STREAK:** N/A

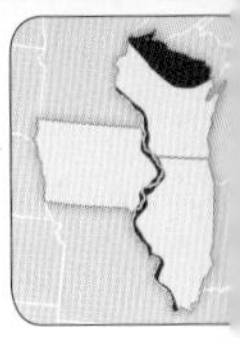
Primary Occurrence

ENVIRONMENT: Lakeshores, rivers, quarries

WHAT TO LOOK FOR: Fine-grained rocks containing coarse, often rectangular, light-colored crystals embedded within

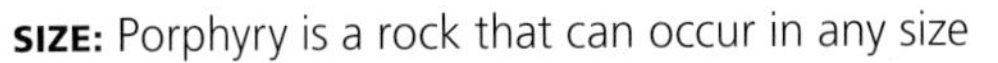

SIZE: Porphyry is a rock that can occur in any size

COLOR: Varies; typically gray or reddish with spots of other colors

OCCURRENCE: Uncommon in Wisconsin; rare in IA and IL

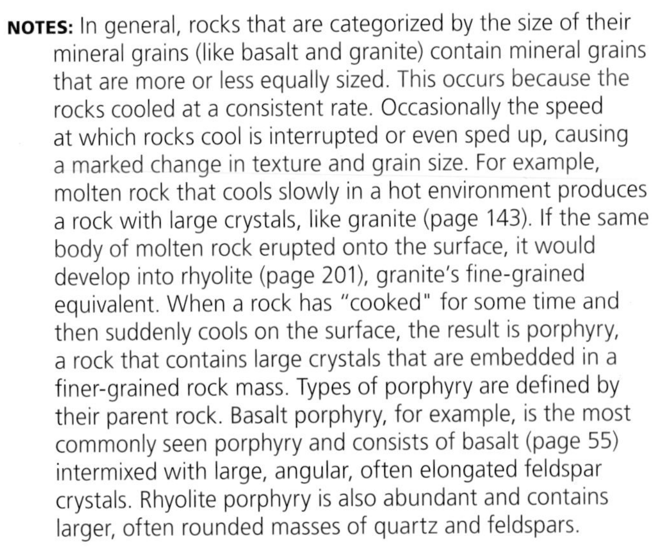

NOTES: In general, rocks that are categorized by the size of their mineral grains (like basalt and granite) contain mineral grains that are more or less equally sized. This occurs because the rocks cooled at a consistent rate. Occasionally the speed at which rocks cool is interrupted or even sped up, causing a marked change in texture and grain size. For example, molten rock that cools slowly in a hot environment produces a rock with large crystals, like granite (page 143). If the same body of molten rock erupted onto the surface, it would develop into rhyolite (page 201), granite's fine-grained equivalent. When a rock has "cooked" for some time and then suddenly cools on the surface, the result is porphyry, a rock that contains large crystals that are embedded in a finer-grained rock mass. Types of porphyry are defined by their parent rock. Basalt porphyry, for example, is the most commonly seen porphyry and consists of basalt (page 55) intermixed with large, angular, often elongated feldspar crystals. Rhyolite porphyry is also abundant and contains larger, often rounded masses of quartz and feldspars.

WHERE TO LOOK: The shores of Lake Superior in northern Wisconsin are best for finding various porphyry, but many rivers, especially large ones like the Mississippi, will also produce rounded samples.

Crust of prehnite crystals on basalt
Water-worn
prehnite
Malachite
Copper
Crude prehnite
Specimen courtesy of Rob Carlson

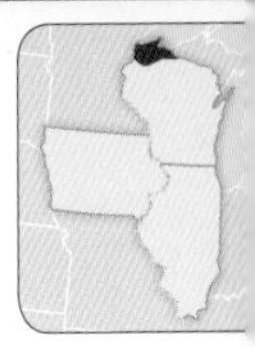

Primary Occurrence

Prehnite

HARDNESS: 6–6.5 **STREAK:** White

ENVIRONMENT: Lakeshores, rivers, outcrops, mines

WHAT TO LOOK FOR: Pale green crusts of hard, lumpy, translucent material lining cavities in dark volcanic rocks

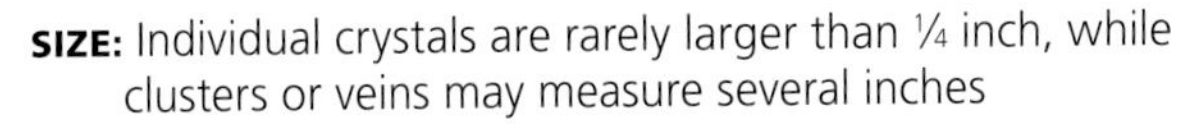

SIZE: Individual crystals are rarely larger than ¼ inch, while clusters or veins may measure several inches

COLOR: Pale green to gray, less commonly colorless or pink

OCCURRENCE: Uncommon in Wisconsin; not found in IA or IL

NOTES: Prehnite, a calcium-, aluminum- and silica-bearing mineral, is found within the dark igneous rocks of northern Wisconsin, particularly inside cracks and vesicles (gas bubbles) in basalt. A fairly distinctive mineral that isn't difficult to identify when well developed, prehnite formed when basalt cavities were filled with hot, mineral-rich water. Typically pale green in color, translucent and always quite hard, prehnite primarily develops as botryoidal (lumpy, grape-like) crystal clusters and crusts lining the cavities it formed within, sometimes filling them completely. Under magnification, you may see that each round prehnite cluster is actually made up of many tiny wedge- or blade-shaped crystals, and when a botryoidal "lump" is broken, its cross section will appear almost fibrous, typically exhibiting a radiating or fan-shaped structure. This appearance when broken, combined with its hardness, is a good way to identify massive specimens, or those with no evident crystal structure. Minerals commonly found associated with prehnite include chlorite, calcite, copper and some zeolites; this will also aid in identification.

WHERE TO LOOK: Prehnite is only found in northern Wisconsin in the vicinity of Lake Superior. Cavities and veins within basalt from Douglas County, particularly near Superior or along the Brule and Amnicon Rivers, have produced nice specimens.

Calcite
Coral fossils
Complex pyrite crystals

Complex pyrite crystals on quartz

Pyrite

HARDNESS: 6–6.5 **STREAK:** Greenish gray

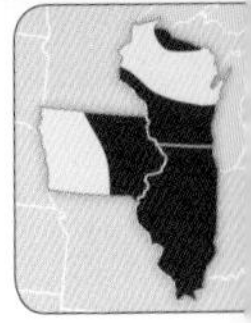

Primary Occurrence

ENVIRONMENT: Outcrops, quarries, mines, fields, road cuts

WHAT TO LOOK FOR: Hard, metallic, brassy yellow masses embedded in rock, or cubic crystals with striated (grooved) sides

SIZE: Crystals tend to be no larger than an inch, but masses or veins can grow to several inches or more in size

COLOR: Brass-yellow to metallic brown

OCCURRENCE: Common

NOTES: Pyrite, one of the world's most common metallic minerals, is a popular collectible in our region, as it is fairly easy to find and identify. Long known as "fool's gold," pyrite is always brassy yellow in color (unless weathered, in which case it can be more brown) and often forms as cubic crystals that are intergrown with each other and have striated (grooved) sides. Rarer crystal shapes are present in the region, too. Examples include octahedrons (a shape resembling two four-sided pyramids placed bottom-to-bottom) and more complex ball-like shapes. Massive or granular varieties are also common in the limestone of the region, but no matter the form it takes, pyrite's color and hardness are distinctive. Chalcopyrite (page 77) is easily mistaken for pyrite but is much softer, while marcasite (page 165) is similar in hardness but typically grayer and forms as flatter crystals. Finally, pyrite is much harder than gold (page 141).

WHERE TO LOOK: Pyrite is ubiquitous throughout the region; limestone in all of southwestern Wisconsin is rich with crystals, as are Iowa and Illinois' famous geodes from along the Mississippi River, especially near Hamilton, IL. Other cavities and veins in sedimentary rocks in Iowa, particularly in quarries in Black Hawk County, have yielded rare octahedral crystals.

Irregularly shaped pyrite "sun disk"

Circular pyrite "sun disk"
Specimen courtesy of Jeff Theroux

Pyrite (continued)

HARDNESS: 6–6.5 **STREAK:** Greenish gray

Primary Occurrence

ENVIRONMENT: Outcrops, quarries, mines, fields, road cuts

WHAT TO LOOK FOR: Hard, metallic, brassy yellow disk-like formations or tiny thread-like crystals in geodes

SIZE: Can be found in sizes up to several inches

COLOR: Brass-yellow to metallic brown

OCCURRENCE: Common to very rare, depending on variety

NOTES: A number of unusual varieties of pyrite are found in the region, and these specimens are even more sought after than the usual cubic and octahedral crystals. The most desired are the peculiar "sun disks" from Sparta, Illinois. Also called "pyrite dollars," these specimens crystallized as flat, circular disk-like masses that formed between layers of shale deep within a coal mine. How they developed is still something of a mystery, but these unusual specimens are no longer obtainable, and if you want to add one to your collection, you need to purchase one. If you do, be aware that they very often turn gray and disintegrate when stored in humid environments. Filiform (thread-like) pyrite is another extremely rare, unique pyrite variety found within the region's geodes (page 131). These tiny, thin, elongated crystals require a microscope to appreciate, and may even go unnoticed until a specimen is examined under magnification. A rusty limonite (page 157) stain may be your first clue of such a find, especially when contrasted on white quartz or calcite.

WHERE TO LOOK: Sparta, Illinois' "sun disks" can no longer be collected but are still widely available for sale. Filiform specimens have been found in geodes from quarries near Hamilton, Illinois, but they are very rare.

Augite in gabbro
Glassy luster
Radiating clusters of aegirine crystals in pegmatite

Pyroxene Group

HARDNESS: 5–6.5 **STREAK:** White to gray-green

Primary Occurrence

ENVIRONMENT: Quarries, rivers, outcrops, road cuts

WHAT TO LOOK FOR: Hard, darkly colored glassy masses, or crystals embedded in coarse-grained rocks, like gabbro or granite

SIZE: Most pyroxene specimens are smaller than ¼ inch

COLOR: Dark brown to black is most common; sometimes green

OCCURRENCE: Common in Wisconsin; rare elsewhere

NOTES: The pyroxenes are a family of "rock-builders" (minerals important to the formation of many types of rock) and are closely related to the amphibole group (page 45). They are very common constituents of igneous rocks, like granite and gabbro, and they are also found in metamorphic rocks. Though many pyroxene minerals exist, only a few are present in the region; the list includes augite, aegirine, diopside and hedenbergite. Augite is the most common, frequently occurring as black, glassy, embedded masses or blocky crystals that are most evident within coarse-grained rocks. These grains or crystals are often nondescript, but all pyroxenes share an identifying trait: when broken, they cleave at nearly ninety-degree angles, making for blocky, step-like breaks. Close observation of the shiny, dark spots in granite or gabbro often illustrate this. Amphiboles exhibit a similar trait but in general tend to have a more fibrous or silky appearance under magnification. Finally, pyroxenes in metamorphic rocks can appear as green needle-like crystals.

WHERE TO LOOK: The pyroxenes are most prominent in igneous rocks, so the lack of igneous rock outcrops in Illinois and Iowa means that pyroxene minerals are primarily a Wisconsin find. Quarries near Wausau have produced aegirine in granite-like rocks, and the basalt and gabbro formations near Lake Superior often reveal embedded augite crystals.

Quartz crystals on quartzite

Quartz crystal

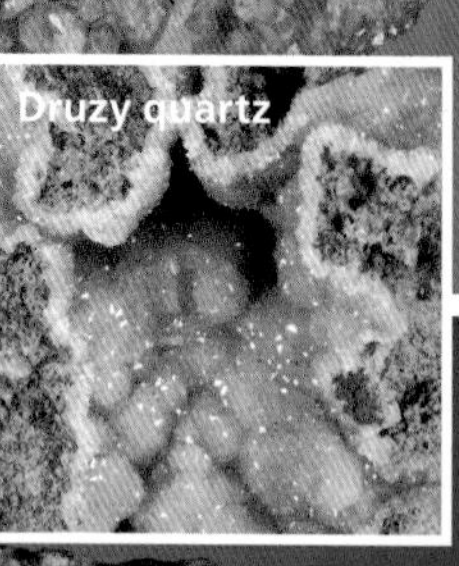

Beach-worn quartz

Quartz (white) in granite

Quartz druse on limestone

Quartz

HARDNESS: 7 **STREAK:** White

Primary Occurrence

ENVIRONMENT: All environments

WHAT TO LOOK FOR: Light-colored, translucent, glassy and very hard six-sided crystals; also found as masses, veins, or river pebbles

SIZE: Crystals can be up to several inches; masses can be any size

COLOR: Colorless to white when pure; often stained yellow to brown, more rarely purple, dark gray, or reddish

OCCURRENCE: Very common

NOTES: Quartz is the most abundant mineral in the earth's crust and the first mineral all new collectors should study. It consists entirely of silica, the silicon- and oxygen-bearing compound that helps form hundreds of other minerals. Most abundant as a component of rocks, it often appears in uninteresting white masses in coarse-grained rocks such as granite or as the microscopic grains that make up chert (page 79). Nonetheless, crystals aren't rare and often appear as elongated hexagonal (six-sided) prisms ending in a point. Crusts or cavity-linings of many tiny intergrown quartz crystals are also common in the region; such crusts are called druzy quartz. White water-worn pebbles and chunky, coarse masses of quartz are found most anywhere, often buried in glacial till (gravel). All quartz is easy to identify, thanks to its distinctive high hardness, translucency, and conchoidal fracture (when struck, circular cracks appear).

WHERE TO LOOK: Quartz is found anywhere throughout the region, especially as worn pebbles. The Mississippi River cuts through areas rich with limestone that yield cavities and geodes lined with quartz. Look in southern Wisconsin near Prairie du Chien, in Iowa near Keokuk, and around Hamilton in Illinois. The Lake Superior shore and granite quarries near Wausau, Wisconsin, also produce fine crystals.

Druzy "strawberry" quartz on chert
Amethyst nodule
Specimen courtesy of
Phil Burgess
Smoky quartz in
basalt
Very fine smoky quartz (gray quartz) crystal

Quartz, Varieties

HARDNESS: 7 **STREAK:** White

Primary Occurrence

ENVIRONMENT: All environments

WHAT TO LOOK FOR: Quartz of visibly different and unique colors and forms

SIZE: Quartz crystals can measure up to an inch or more; masses can be almost any size

COLOR: Red to brown, purple, gray to black, or multicolored, depending on variety

OCCURRENCE: Uncommon

NOTES: Quartz is so abundant in nearly every geological environment that a huge number of variables can affect its development, resulting in quartz with different colors and growth habits. A type of formation called druse is particularly prevalent in the region, appearing as a coating of countless tiny quartz crystals on rock surfaces or in cavities. Druzy quartz often exhibits a "sparkly" appearance because each tiny lustrous crystal reflects light in different directions. Agates (page 37) are perhaps one of the most famous quartz variations; they consist of microscopic quartz arranged into layers; chert (page 79) and jasper (page 151) are similar. Color variations in quartz are also popular with collectors and are derived from impurities within the crystals. Red quartz, sometimes called "cinnamon" or "strawberry" quartz, gets its color from abundant iron-bearing mineral impurities within it; purple quartz, called amethyst, gets its color from naturally irradiated iron and aluminum impurities.

WHERE TO LOOK: Red quartz druse is abundant in southwestern Wisconsin, near the Mississippi River and is often associated with stromatolite fossils (page 119). Amethyst pebbles may be found, especially near Lake Superior, and agates frequently turn up along rivers and lakes in all three states.

Texture detail
Quartzite fragments
Faint layering
Quartzite fragments

Quartzite

HARDNESS: ~7 **STREAK:** N/A

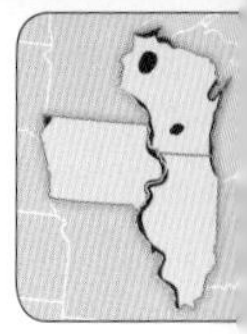

Primary Occurrence

ENVIRONMENT: All environments

WHAT TO LOOK FOR: Very hard, grainy rock with the general appearance and characteristics of quartz

SIZE: As a rock, quartzite can be found in any size

COLOR: White to gray finds are common; often stained red to brown, yellow, or green to black; occasionally banded

OCCURRENCE: Very common

NOTES: Quartzite is a metamorphic rock that forms when sandstone (page 199) is altered; quartzite can form in two different ways. Sandstone can be compressed and heated, causing the sand grains within it to fuse together; quartzite can also form when sandstone is impregnated by silica (quartz) solutions, which then crystallize into quartz between the grains. Because sand consists primarily of grains of quartz, quartzite is composed almost entirely of quartz. It is therefore very hard, tough and water-resistant and is often found in glacial till and along rivers as rounded stones. It is generally light colored, though it's often stained darker by impurities, and it may be translucent. It is very easy to confuse with typical quartz (page 195), but quartz is always more translucent and appears glassier when broken. Quarzite also has more of a grainy, flaky texture than quartz, especially under magnification. Chert (page 79) may also appear similar, but it is always more opaque and often waxier in texture.

WHERE TO LOOK: The much-studied Baraboo Hills near Baraboo, Wisconsin, are composed largely of quartzite, and a tiny portion of the far northwestern corner of Lyon County in Iowa is underlain by quarzite. In Illinois and elsewhere in the region, quartzite is common in glacial till and river gravel.

Beach-worn rhyolite
Flow-banding
Texture detail
Rough rhyolite
Rhyolite porphyry

Rhyolite

HARDNESS: 6–6.5 **STREAK:** N/A

Primary Occurrence

ENVIRONMENT: Lakeshores, rivers, outcrops, road cuts, quarries

WHAT TO LOOK FOR: Hard, dense, fine-grained rock, typically light colored or reddish and often containing gas bubbles

SIZE: As a rock, rhyolite can be found in any size

COLOR: Varies greatly; typically gray to tan, also commonly brown to reddish; occasionally with bands of other colors

OCCURRENCE: Common in Wisconsin; uncommon elsewhere

NOTES: Granite (page 143) forms when a body of molten rock cools slowly, deep within the earth; if that same molten rock is instead erupted onto the earth's surface, it cools very rapidly, resulting in rhyolite. Because it cooled quickly, the minerals in rhyolite—quartz, potassium feldspars, amphiboles and micas, among others—were allowed little time to crystallize, so the rock is very fine grained and generally evenly colored, much like basalt (page 55). Basalt, however, contains more dark-colored minerals, while rhyolite has a much higher silica (quartz material) content; this means that rhyolite is slightly harder and lighter in color than basalt. Rhyolite lava is also more viscous (thicker) than basalt lava, so although it did cool rapidly, rhyolite lava held its heat longer than basalt and its mineral grains are often slightly larger. Rhyolite is generally grayish, but reddish hues are very common as well, and vesicles (gas bubbles) and bands of color (caused by the motion of the rock while molten) are abundant. But you'll primarily find rhyolite only in Wisconsin.

WHERE TO LOOK: Much of the bedrock in north-central Wisconsin, as well as in the counties bordering Lake Superior, is rhyolite. Douglas and Taylor counties are good places to look. In Iowa and Illinois, you'll only find rhyolite in gravel.

Sandstone
Layering
Algae (green
on sandstone
Texture detail
Iron-stained
sandstone
Sandstone concretions
Water-worn sandstone

Sandstone

HARDNESS: N/A **STREAK:** N/A

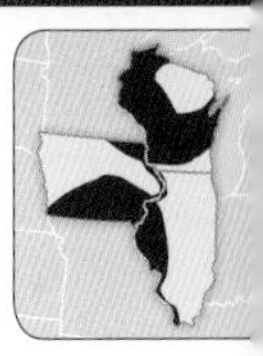

Primary Occurrence

ENVIRONMENT: All environments

WHAT TO LOOK FOR: Rocks with a very gritty texture and granular nature, often easily picked apart with your fingernails

SIZE: As a rock, sandstone can be found in any size

COLOR: Varies greatly; generally red to brown or yellow, sometimes layered, and less commonly gray to green

OCCURRENCE: Very common

NOTES: Like most sedimentary rocks, sandstone gets its name from the particles it consists of. Sand is a sediment we're all familiar with; it consists of detrital particles no larger than 1⁄12 of an inch and is primarily made up of quartz (page 195), along with minor amounts of other worn-down mineral grains. Sand turned to sandstone when it settled into beds at the bottoms of ancient lakes and seas, where it was later compressed by pressure from above; other minerals, particularly clay and calcite, formed between the grains, cementing them together. Sandstone therefore has a rough, gritty texture, and individual grains of sand are often easy to separate with only your fingernails (depending on the level of cementation of the particular sandstone). Under magnification, quartz grains are easily visible, and layers of varying coloration are common. For these reasons, sandstone is perhaps the easiest sedimentary rock to identify.

WHERE TO LOOK: Sandstone is common throughout the region, due to the ancient seas that once flooded the area. Much of eastern Iowa, southern Wisconsin, and north-central Illinois is underlain by sandstone bedrock, though most of it is buried. You will find lots of sandstone in glacial gravel, but the Bayfield Peninsula in Lake Superior provides some of the region's best views of sandstone outcrops and cliffs.

Shale
Shale texture detail
Separated layers
Layers in slate
Slate

Shale/Slate

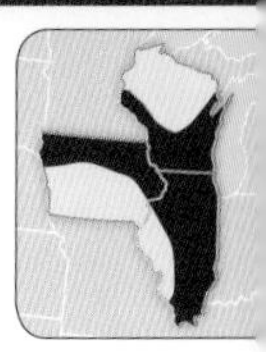
Primary Occurrence

HARDNESS: <5.5 **STREAK:** N/A

ENVIRONMENT: All environments

WHAT TO LOOK FOR: Soft, fine-grained rocks in flat, layered, sheet-like formations; layers can be separated with a knife

SIZE: As rocks, shale and slate can be found in any size

COLOR: Shale is tan to brown, gray to black; slate is dark gray

OCCURRENCE: Shale is common; slate is uncommon

NOTES: The inland seas that repeatedly inundated the region left behind many kinds of sedimentary rocks. Shale, consisting of layers of compacted, solidified mud, is among the most common and formed at the bottom of very still bodies of water. Shale consists of microscopic grains of weathered rocks and minerals, particularly micas, clays and quartz, which were periodically deposited and hardened into layers. As a result, shale is a soft rock and can easily be scratched with a knife. It's easy to identify thanks to its layering and habit of softening when soaked in water. In addition, shale's layers are often easily separated with a knife blade and come apart in distinct thin, flat layers; fossils are often found between layers. One of the only rocks you'll confuse with shale is mudstone (page 177), which is, in essence, unlayered shale. Slate is a rock formed when shale is subjected to intense pressure and/or heat, compacting and hardening the shale into a much harder, darker, finer-layered rock. Slate's layers are not as easily separated and are more brittle, and slate doesn't soften when soaked in water, as shale does.

WHERE TO LOOK: Throughout the region, you won't have trouble finding shale, as it underlies huge amounts of Iowa and Illinois, as well as southern Wisconsin; look along rivers. Slate is rarer and is primarily found in northern Wisconsin, particularly in Bayfield, Ashland and Florence counties.

Bladed siderite crystals in geode
Specimen courtesy of Phil Burgess

Siderite

HARDNESS: 3.5–4 **STREAK:** White

Primary Occurrence

ENVIRONMENT: Mines, quarries, outcrops, road cuts

WHAT TO LOOK FOR: Small, brown, pearly rhombohedral (shaped like a leaning cube) crystals or masses with step-like surfaces

SIZE: Siderite crystals are typically smaller than ¼ inch, while masses can measure up to palm-sized or rarely larger

COLOR: Tan to brown, dark brown, cream-colored to yellow

OCCURRENCE: Uncommon

NOTES: Consisting of iron carbonate, siderite is a close cousin to calcite (page 65) and dolomite (page 97) and is found in almost any iron-rich area throughout the region. It primarily forms in sedimentary environments, often as masses or veins within shale or limestone, but it may be found in other kinds of rock as well. These massive occurrences of siderite are often unattractive, easily overlooked and may not be easily identified. Crystals are more collectible and take the form of rhombohedrons (a shape like a leaning or skewed cube), or sometimes occur as a more flattened shape resembling a curved blade. However it appears, siderite typically exhibits a pearly luster and is generally tan to brown in color. Though similar, calcite is slightly softer, while dolomite does not become slightly magnetic when heated in a flame, as siderite will. Siderite also has perfect rhombohedral cleavage, so specimens will break into rhombohedral shapes when struck. Finally, many fossils, including the concretions (page 89) from Mazon Creek, Illinois, are largely siderite.

WHERE TO LOOK: Crystals have been found in geodes from Keokuk and Van Buren counties, Iowa, and in Brown County, Illinois. In Wisconsin, exposures of rock in the Gogebic Iron Range of Ashland and Iron counties may yield veins or masses of siderite, particularly near Montreal and Hurley.

Sillimanite (brownish) in schist
Fibrous crystals
Fine fibrous crystals
Fine fibrous crystals

Sillimanite

HARDNESS: 6.5–7.5 **STREAK:** White

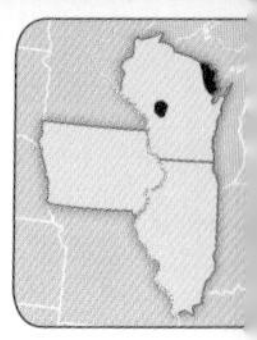

Primary Occurrence

ENVIRONMENT: Mines, outcrops

WHAT TO LOOK FOR: Brown, hard fibrous material in metamorphic rocks like schist

SIZE: Most crystals are shorter than ¼ inch; masses may be an inch or more

COLOR: Brown to yellow-brown, gray

OCCURRENCE: Rare in Wisconsin; not found in Iowa or Illinois

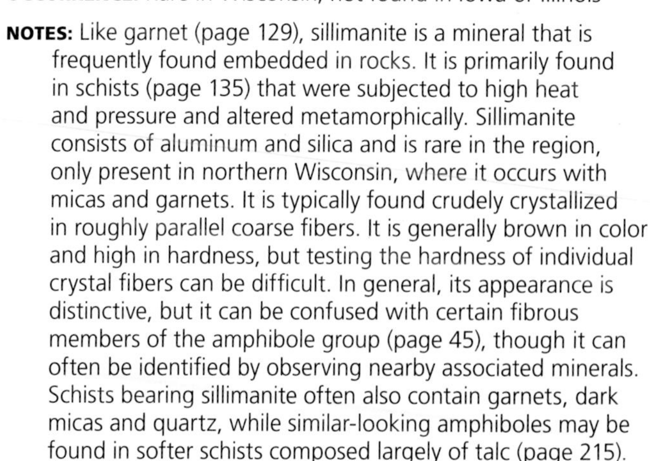

NOTES: Like garnet (page 129), sillimanite is a mineral that is frequently found embedded in rocks. It is primarily found in schists (page 135) that were subjected to high heat and pressure and altered metamorphically. Sillimanite consists of aluminum and silica and is rare in the region, only present in northern Wisconsin, where it occurs with micas and garnets. It is typically found crudely crystallized in roughly parallel coarse fibers. It is generally brown in color and high in hardness, but testing the hardness of individual crystal fibers can be difficult. In general, its appearance is distinctive, but it can be confused with certain fibrous members of the amphibole group (page 45), though it can often be identified by observing nearby associated minerals. Schists bearing sillimanite often also contain garnets, dark micas and quartz, while similar-looking amphiboles may be found in softer schists composed largely of talc (page 215).

WHERE TO LOOK: Most localities for sillimanite in Wisconsin are in the northeast along the Michigan border, including near Thunder Mountain. It is also found in metamorphic rocks in a few scattered outcrops and mines. Look near Black River Falls and Lake Wazee (a former open-pit iron mine), though much of that area is protected.

Fine sphalerite crystals on dolomite
Internal reflections
Crude intergrown crystals
Galena
Fine sphalerite crystals

Sphalerite

HARDNESS: 3.5–4 **STREAK:** Light brown

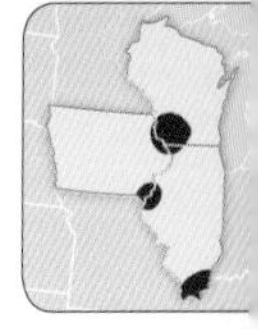

Primary Occurrence

ENVIRONMENT: Quarries, mines, outcrops

WHAT TO LOOK FOR: Lustrous, typically dark-colored crystals with many triangular faces or masses occurring with chalcopyrite

SIZE: Crystals are generally smaller than an inch or two; masses can become quite large, up to several inches

COLOR: Reddish brown to black, more rarely yellow to greenish

OCCURRENCE: Uncommon

NOTES: Sphalerite is the primary ore of zinc, and it is an extremely important mineral in the region, both economically as well as from a collector's viewpoint. It typically formed in our area within limestone (page 155) while the limestone itself was forming; as the aragonite shells of aquatic organisms and coral began to convert into masses of calcite (the process by which most of the region's limestone developed), impurities including lead and zinc were freed. These impurities then reacted with sulfur that was released from coal and oil, and crystallized. For this reason, sphalerite is very commonly found with chalcopyrite (page 77) as well as galena (page 127), with which it could possibly be confused (though galena is much softer). Sphalerite is typically reddish brown to black and may appear nearly metallic and opaque, but broken or thin specimens will show some translucency and bright internal reflections. Crystals are often complex in shape but frequently exhibit triangular faces and points.

WHERE TO LOOK: Grant, Iowa and Lafayette counties in Wisconsin are very lucrative, especially near Mineral Point and Cuba City. In Iowa, Scott and Lee counties have produced wonderful crystals, even within geodes near Keokuk. Hardin County in Illinois is famous for fine specimens, but the mines that produced them are off-limits to casual collectors.

Strontianite crystal clusters

Crust of tiny strontianite crystals arranged into radial groups

Strontianite

HARDNESS: 3.5 **STREAK:** White

Primary Occurrence

ENVIRONMENT: Mines

WHAT TO LOOK FOR: Light-colored needle-like crystal groups arranged into balls or divergent "sprays"

SIZE: Crystals tend to be shorter than an inch, but groups can measure several inches

COLOR: Colorless to white or gray, yellowish

OCCURRENCE: Very rare in Illinois and Wisconsin; not found in Iowa

NOTES: One of the region's most elusive minerals, strontianite was named not for its strontium content but for the Scottish town where it (and strontium) were discovered. Related to aragonite, strontianite forms primarily within limestone, but despite the large amount of sedimentary rock in the region, it remains a rare mineral. When crystallized, it develops as thin, delicate needles that are typically arranged into semi-spherical groupings or divergent "spray"-like clusters. Sometimes the individual crystals are distinctly curved, but not always, especially in the southern Illinois mines where very fine specimens have been found. Identifying specimens can be difficult because of their great resemblance to calcite (page 65) and aragonite (page 49), but there are a few ways of telling them apart. Calcite is slightly softer and won't easily scratch brass, while strontianite will. Aragonite is harder but is more common, slightly harder and often forms as coarser crystals.

WHERE TO LOOK: This rare mineral has been found in quarries north of Appleton, Wisconsin, but some of the world's finest specimens originated from the Cave-in-Rock mining district in Hardin County, Illinois. Unfortunately, those mines are off-limits to collectors without permission, though specimens can often be found for sale.

Mass of greenish talc
Bright greasy luster
Talc mass on schist
Dark steatite mass

Talc/Steatite

HARDNESS: 1 **STREAK:** White

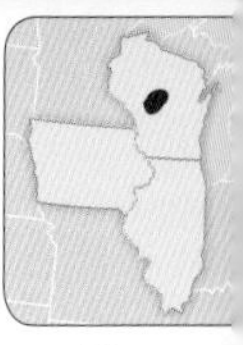

Primary Occurrence

ENVIRONMENT: Quarries, mines, outcrops

WHAT TO LOOK FOR: Extremely soft, flaky mineral in metamorphic rocks; or compact, finely layered and very soft green rock

SIZE: Masses of talc or steatite can be several feet in size

COLOR: Talc is often pale green to green, or white to gray; steatite is dark green to black

OCCURRENCE: Rare in Wisconsin; not found in IA or IL

NOTES: On the Mohs scale of mineral hardness, talc is the softest mineral. It is a magnesium-rich mineral that forms when olivine, pyroxenes, amphiboles or serpentines are extensively altered, which occurred in this region when dark rocks were metamorphosed. As a result, talc is most often found not as crystals but as scaly masses or veins in schists in northern Wisconsin, particularly in iron-rich areas. Typically light in color and often greenish, talc is very easy to identify thanks to its extremely low hardness, flaky habit, bright "greasy" luster, and "soapy" or even slippery feel. But talc is quite rare in the region and found in just a handful of areas. Steatite, also known as soapstone, is a unique variety of schist that consists almost entirely of talc, along with some amphiboles and chlorite. It is quite dark in color and, like most schists, it is finely layered but very soft (typically under 2 in hardness) and easy to cut with a knife; it has long been popular for carving. Steatite is rare in Wisconsin, however.

WHERE TO LOOK: Many of Wisconsin's talc localities are in Wood County, and the area around Milladore has produced both talc and carvable steatite. In Jackson County, the vicinity of Lake Wazee (a former iron mine) produces talc-bearing schists, but much of that area is a protected park.

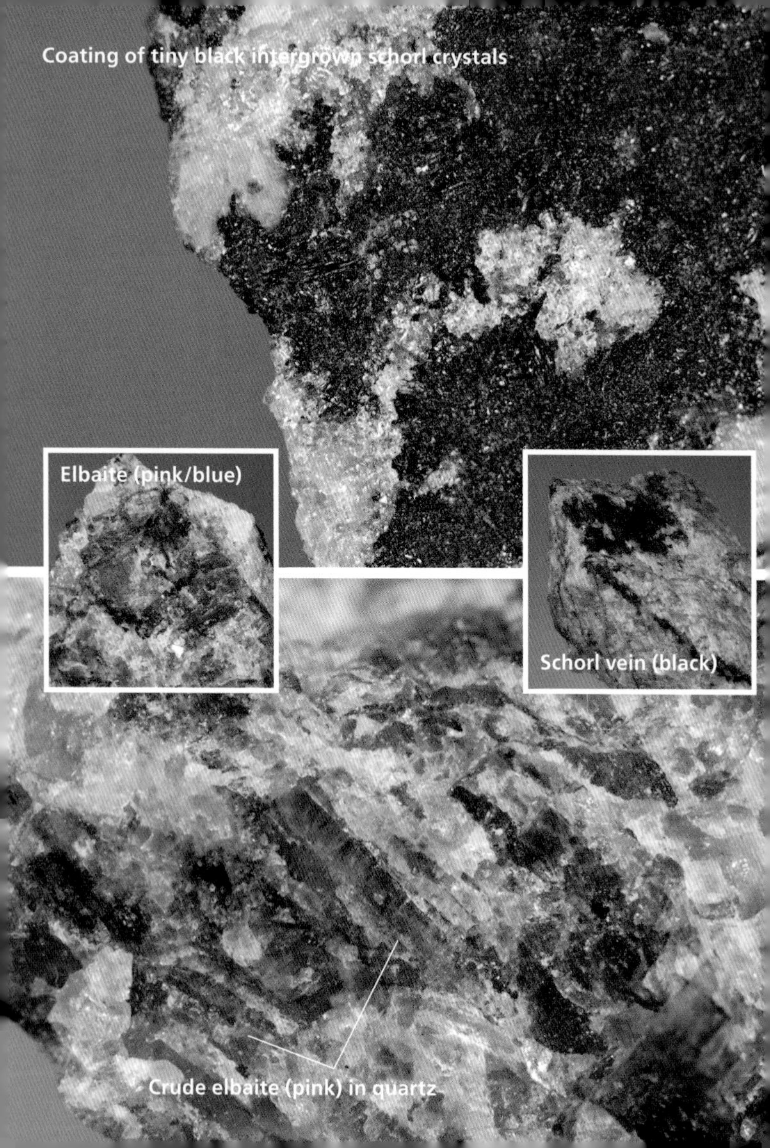
Coating of tiny black intergrown schorl crystals
Elbaite (pink/blue)
Schorl vein (black)
Crude elbaite (pink) in quartz

Tourmaline Group

HARDNESS: 7–7.5 **STREAK:** White

Primary Occurrence

ENVIRONMENT: Quarries, outcrops

WHAT TO LOOK FOR: Very hard, slender, elongated crystals with striated (grooved) sides embedded in rocks like granite

SIZE: Crystals are generally shorter than an inch; masses or veins may measure several inches

COLOR: Black common; rarely blue to pink

OCCURRENCE: Rare in Wisconsin; very rare in Iowa and Illinois

NOTES: Several members of the popular tourmaline group of minerals exist, but only two are prevalent in the region: schorl and elbaite. Schorl, the more common of the two, is opaque, black and has a bright glassy luster, forming as long, slender needle-like crystals with striated (grooved) sides when well developed. If large enough to be visible, a crystal will have a more-or-less triangular cross section; when combined with its high hardness, this makes schorl easy to identify. Poorer examples appear as crusts or veins of tiny intergrown needles, sometimes arranged radially, and won't give you an accurate hardness. Elbaite shares all the same traits as schorl but is generally more colorful in shades of pink to blue. Both are easiest to find in granite or pegmatite (very coarse granite) outcrops, where they are often embedded in quartz or feldspar. Some amphiboles (page 45) may look similar to schorl but tourmalines are harder.

WHERE TO LOOK: With the exception of tiny crystals in sand or possible tourmaline crystals embedded within the boulders spread throughout the region by glaciers, tourmaline minerals will only be found in Wisconsin. Florence, Forest and Marinette counties have produced lots of specimens from pegmatites and granite near Alvin, Pembine, Aurora and other quarries and outcrops near the Michigan border.

Water-worn unakite pebbles
Epidote-rich pebble
Epidote (greenish)
Feldspars (orange)

Unakite

HARDNESS: N/A **STREAK:** N/A

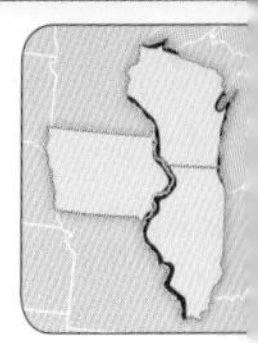

Primary Occurrence

ENVIRONMENT: Lakeshores, rivers

WHAT TO LOOK FOR: Green and orange granite-like pebbles, typically found in rivers or on lakeshores

SIZE: Unakite is a rock that can occur in any size, though samples are typically smaller than a fist

COLOR: Varies; primarily green and orange, pink, white or tan

OCCURRENCE: Uncommon; more common in Wisconsin

NOTES: As the glaciers of the past ice ages pushed southward from Canada into Wisconsin, Iowa and Illinois, they carried enormous amounts of pulverized rock and gravel with them, scattering it throughout the region. For this reason it is possible to find samples of rocks that didn't form in the region but instead originated from farther north; unakite is a prime example of this. Often conspicuous with its green and orange or pink color combination, unakite is actually a variety of granite (page 143) that long ago was affected by hot volcanic water which altered some of its feldspars (page 101) and turned them into epidote (page 99), with its characteristic yellowish green coloration. Other varieties of feldspar remained intact and are visible as the orange, pink or tan spots. Quartz (page 195) is also typically present as gray or colorless spots. In most cases, identifying unakite is easy, as its coloration is distinctive and it will only be found as loose pebbles in river gravel.

WHERE TO LOOK: You'll find unakite most frequently on the shores of Lake Superior and other northern Wisconsin shorelines, but any river in southern Wisconsin, northeastern Iowa, and most of Illinois will potentially yield samples. In fact, virtually any area that saw glaciation may produce specimens from within local gravel.

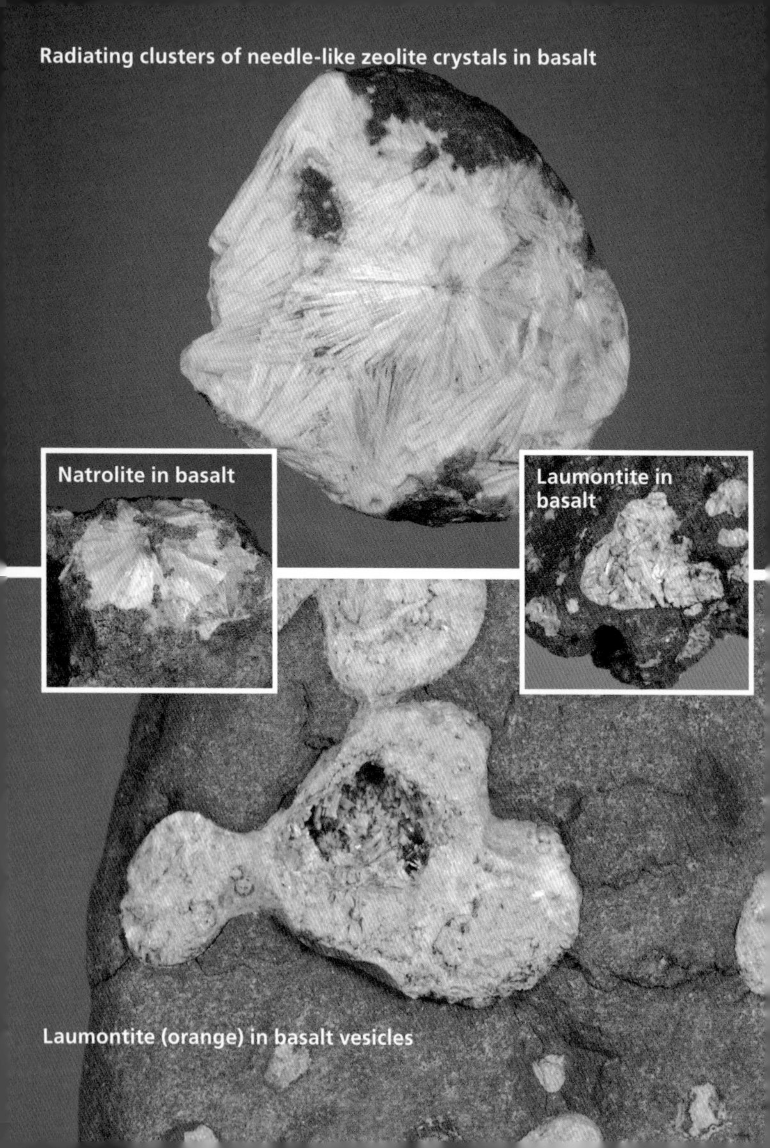

Radiating clusters of needle-like zeolite crystals in basalt

Natrolite in basalt

Laumontite in basalt

Laumontite (orange) in basalt vesicles

Zeolite Group

HARDNESS: 3.5–5.5 **STREAK:** Colorless to white

Primary Occurrence

ENVIRONMENT: Lakeshore, outcrops, road cuts

WHAT TO LOOK FOR: Small, brittle, often fibrous light-colored crystals filling cavities in dark volcanic rocks

SIZE: Individual crystals tend to be very small, less than ¼ inch; crystal clusters may be up to an inch or more

COLOR: Varies depending on species; typically colorless to white or gray, pink to orange, or tan to brown

OCCURRENCE: Uncommon in Wisconsin; not found in IA or IL

NOTES: The zeolites are a large family of minerals that form primarily within vesicles (gas bubbles) and other cavities in Wisconsin's basalt (page 55) and, to a lesser extent, gabbro. They form when the rock is affected by alkaline groundwater. They are chemically complex minerals and exhibit very different crystal shapes depending on species. Natrolite is one of the most abundant types in Wisconsin and takes the form of tiny needle-like crystals, often intergrown in silky lustered masses and arranged into divergent "sprays." Stilbite is less common and tends to appear as flat, plate-like, pearly lustered crystals on cavity walls. Analcime, a very uncommon zeolite in Wisconsin, appears as glassy, faceted, ball-like crystals. Laumontite is more common and forms elongated, rectangular crystals that are often orange and occur in crumbly masses. In all of these cases, identifying Wisconsin's zeolites can be difficult. When looking through cavities in basalt, watch for tiny, white, brittle crystals with a pearly or silky luster, often intergrown with calcite (page 65), chlorite, clay or quartz.

WHERE TO LOOK: Of the three states, only Wisconsin yields zeolites, almost exclusively in basalt formations on or near Lake Superior's shores. Douglas and Bayfield counties are lucrative; look along the Amnicon River southeast of Superior.

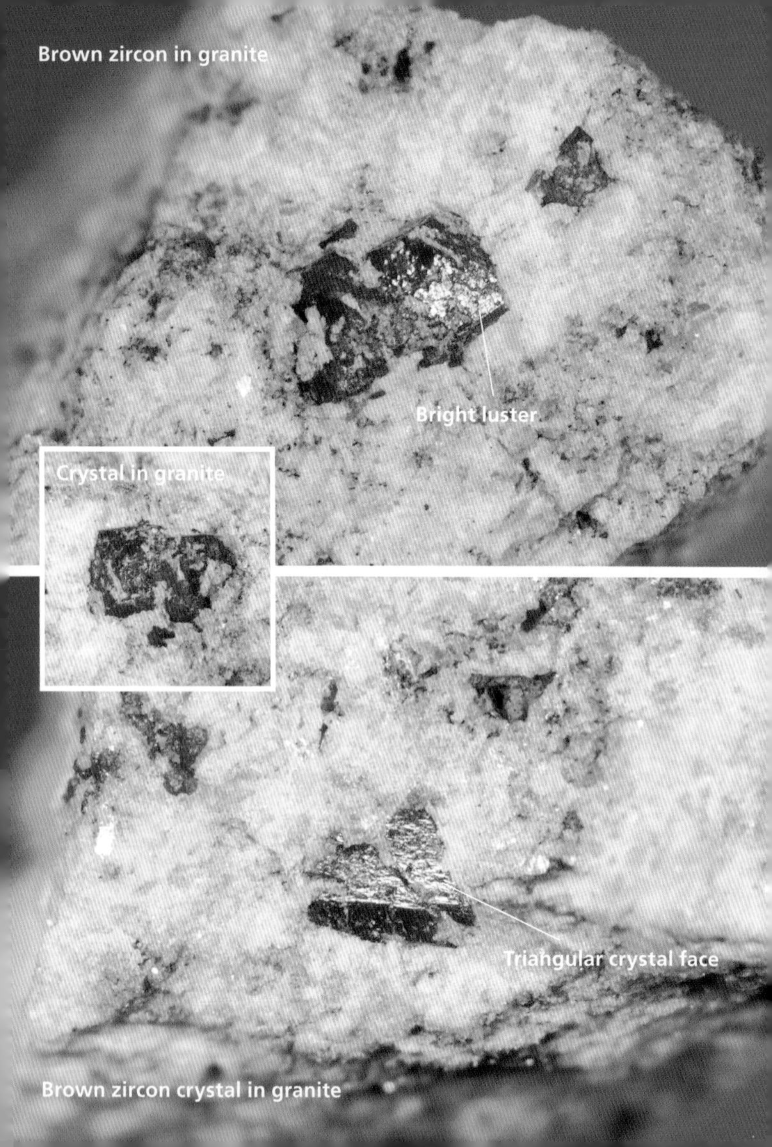
Brown zircon in granite
Bright luster
Crystal in granite
Triangular crystal face
Brown zircon crystal in granite

Zircon

HARDNESS: 7.5 **STREAK:** Colorless

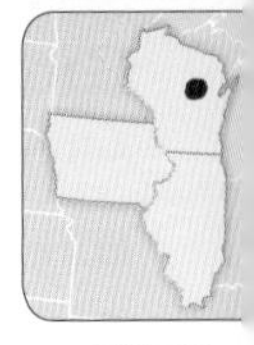

Primary Occurrence

ENVIRONMENT: Quarries, mines, outcrops, rivers, road cuts

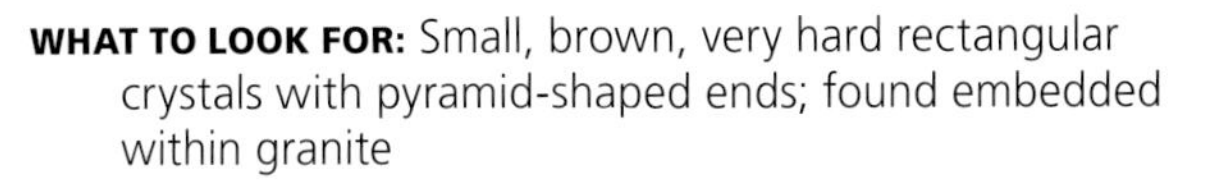

WHAT TO LOOK FOR: Small, brown, very hard rectangular crystals with pyramid-shaped ends; found embedded within granite

SIZE: Most zircon crystals are very small, no larger than ⅛ inch

COLOR: Gray to brown; more rarely white, pink to red, or purple

OCCURRENCE: Rare in Wisconsin; very rare in Iowa and Illinois

NOTES: Zircon, a mineral long used in jewelry, is technically an abundant and widespread mineral, but not in the sense that you may hope. Many kinds of rocks, particularly granite and sandstone, contain countless tiny inconspicuous zircon grains, many of which are nearly microscopic. Unfortunately, fine crystals of any size are rare, and you'll only find them in Wisconsin's granite. When ideally formed, zircons appear as elongated rectangular prisms that are tipped with a four-sided pyramid on each end. Combined with their usual square cross section, this is a highly distinctive crystal form. And though hardness is difficult to check on such small crystals, zircons are notably very hard. Unfortunately, color and luster are too inconsistent to aid in identification; while most are brightly lustrous and pale brown in color, others can vary significantly. To find zircons, break granite and use magnification to inspect the small dark spots and look for shiny triangular shapes. It should also be noted that many zircons are radioactive, though they are generally so tiny that they do not pose a health risk.

WHERE TO LOOK: With the exception of tiny grains in sandstone and quartzite, collectible zircons are only found in Wisconsin in granite quarries near Wausau, especially those just south of Rib Mountain; be aware that most are private property.

Glossary

AGGREGATE: An accumulation or mass of crystals

ALKALINE: Describes substances containing alkali elements, such as calcium, sodium and potassium; having the opposite properties of acids

ALTER: Chemical changes within a rock or mineral due to the addition of mineral solutions

AMPHIBOLE: A large group of important rock-forming minerals commonly found in granite and similar rocks

ASSOCIATED: Minerals that often occur together due to similar chemical traits

BAND: An easily identified layer within a mineral

BED: A large, flat mass of rock, generally sedimentary

BOTRYOIDAL: Crusts of a mineral that formed in rounded masses; said to resemble a bunch of grapes

BRECCIA: A coarse-grained rock composed of broken angular rock fragments that have been solidified together

CHALCEDONY: A massive, microcrystalline variety of quartz

CLEAVAGE: The property of a mineral to break along the planes of its crystal structure, which reflects its internal crystal shape; referred to in terms of shape or angles

COMPACT: Dense, tightly formed rocks or minerals

CONCENTRIC: Circular, ringed bands that share the same center, with larger rings encompassing smaller rings

CONCHOIDAL: A circular shape, often resembling a half-moon; generally refers to fracture shape

CRATON: A very old block of the earth's crust, typically composed of granite, which forms the core of a continent

CRUST: The rigid outermost layer of the earth

CRYSTAL: A solid body with a repeating atomic structure formed when an element or chemical compound solidifies

CUBIC: A box-like structure with sides of an equal size

DEHYDRATE: To lose water contained within

DETRITUS: Debris, especially plant matter

DRUSE: A coating of small crystals on the surface of another rock or mineral

DULL: A mineral that is poorly reflective

EARTHY: Resembling soil; dull luster and rough texture

EFFERVESCE: When a mineral placed in an acid gives off bubbles caused by the mineral dissolving

ERUPTION: The ejection of volcanic materials (lava, ash, etc.) onto the earth's surface

FACE: A typically smooth surface of a crystal derived from a mineral's growth structure

FELDSPAR: An extremely common and diverse group of light-colored minerals that are most prevalent within rocks and make up the majority of the earth's crust

FIBROUS: Fine, rod-like crystals that resemble cloth fibers

FLUORESCENCE: The property of a mineral to give off visible light when exposed to ultraviolet light radiation

FRACTURE: The way a mineral breaks or cracks when struck, often referred to in terms of shape or angles

GEODE: A hollow rock or mineral formation, typically exhibiting a very round, ball-like external shape and interior walls lined with minerals, namely quartz and calcite

GLASSY: A mineral with a reflectivity similar to window glass; also known as "vitreous luster"

GNEISS: A rock that has been metamorphosed so that some of its minerals are aligned in parallel bands

GRANITIC: Pertaining to granite or granite-like rocks

GRANULAR: A texture or appearance of rocks or minerals that consist of grains or particles

HEXAGONAL: A six-sided structure

HOST: A rock or mineral on or in which other rocks and minerals occur

HYDROUS: Containing water

IGNEOUS ROCK: Rock resulting from the cooling and solidification of molten rock material, such as magma or lava

IMPURITY: A foreign mineral within a host mineral that often changes properties of the host, particularly color

INCLUSION: A mineral that is encased or impressed into a host mineral

IRIDESCENCE: When a mineral exhibits a rainbow-like play of color, often only at certain angles

LAVA: Molten rock that has reached the earth's surface

LUSTER: The way in which a mineral reflects light off of its surface, described by its intensity

MAGMA: Molten rock that remains deep within the earth

MASSIVE: Mineral specimens found not as individual crystals but rather as solid, compact concentrations; rocks are often described as massive; in geology, "massive" is rarely used in reference to size

MATRIX: The rock in which a mineral forms

METAMORPHIC ROCK: Rock derived from the alteration of existing igneous or sedimentary rock through the forces of heat and pressure

METAMORPHOSED: A rock or mineral that has already undergone metamorphosis

MICA: A large group of minerals that occur as thin flakes arranged into layered aggregates resembling a book

MICROCRYSTALLINE: Crystal structure too small to see with the naked eye

MINERAL: A naturally occurring chemical compound or native element that solidifies with a definite internal crystal structure

NATIVE ELEMENT: An element found naturally uncombined with any other elements, e.g. copper

NODULE: A rounded mass consisting of a mineral, generally formed within a vesicle or other cavity

OCTAHEDRAL: A structure with eight-faces, resembling two pyramids placed base-to-base

OPAQUE: Material that lets no light through

ORE: Rocks or minerals from which metals can be extracted

OXIDATION: The process of a metal or mineral combining with oxygen, which can produce new colors or minerals

PEARLY: A mineral with reflectivity resembling that of a pearl

PEGMATITE: The most coarsely grained portion of a granite formation, compose of large, interlocking crystals. Slow cooling, usually in conjunction with water, allowed the minerals within the rock to crystallize to large sizes; often contains rare minerals

PLACER: Deposit of sand containing dense, heavy mineral grains at the bottom of a river or a lake

PRISMATIC: Crystals with a length greater than their width

PSEUDOMORPH: When one mineral replaces another but retains the outward appearance of the initial mineral

PYROXENE: A group of hard, dark, rock-building minerals that make up many dark-colored rocks like basalt or gabbro

RADIATING: Crystal aggregates growing outward from a central point, often resembling the shape of a paper fan

RHOMBOHEDRON: A six-sided shape resembling a leaning or skewed cube

ROCK: A massive aggregate of mineral grains

ROCK-BUILDER: Refers to a mineral important in rock creation

SCHILLER: A mineral that exhibits internal reflections or "flashes" from within its structure when rotated in bright light; often showing an interplay of white, yellow or blue

SCHIST: A rock that has been metamorphosed so that most of its minerals have been concentrated and arranged into parallel layers

SEDIMENT: Fine particles of rocks or minerals deposited by water or wind, e.g. sand

SEDIMENTARY ROCK: Rock derived from sediment being cemented together

SILICA: Silicon dioxide; forms quartz when pure and crystallized, and contributes to thousands of minerals

SPECIFIC GRAVITY: The ratio of the density of a given solid or liquid to the density of water when the same amount of each is used, e.g. the specific gravity of copper is approximately 8.9, meaning that a sample of copper is about 8.9 times heavier than the same volume of water

SPECIMEN: A sample of a rock or mineral

STALACTITE: A cone-shaped mineral deposit grown downward from the roof of a cavity; sometimes described as icicle-shaped; formations in this shape are said to be stalactitic

STRIATED: Parallel grooves in the surface of a mineral

TABULAR: A crystal structure in which one dimension is notably shorter than the others, resulting in flat, plate-like shapes

TARNISH: A thin coating on the surface of a metal, often differently colored than the metal itself (see *oxidation*)

TERRANE: A fragment of the earth's crust that has collided with and embedded into another landmass but retains its own unique geology

TRANSLUCENT: A material that lets some light through

TRANSPARENT: A material that lets enough light through that one can see what lies on the other side

TWIN: Two or more crystals intergrown within or through each other

VEIN: A mineral, particularly a metal, that has filled a crack or similar opening in a host rock or mineral

VESICLE/VESICULAR: A cavity created in an igneous rock by a gas bubble trapped when the rock solidified; a rock containing vesicles is said to be vesicular

VOLCANO: An opening, or vent, in the earth's surface that allows volcanic material such as lava and ash to erupt

VUG: A small cavity within a rock or mineral that can become lined with different mineral crystals

WAXY: A mineral with a reflectivity resembling that of wax

ZEOLITE: A group of similar minerals with very complex chemical structures that include elements such as sodium, calcium and aluminum combined with silica and water; zeolites typically form within cavities in basalt as it is affected by mineral-bearing alkaline groundwater

Wisconsin Museums and Rock Shops

UNIVERSITY OF WISCONSIN GEOLOGY MUSEUM
1215 West Dayton Street
Madison, WI 53706
608-262-2399
www.geology.wisc.edu/~museum/index.html

WEIS EARTH SCIENCE MUSEUM
University of Wisconsin–Fox Valley
1478 Midway Road
Menasha, WI 54952
920-832-2925
www.uwfox.uwc.edu/wesm/index.html

THOMAS A. GREENE GEOLOGICAL MUSEUM
University of Wisconsin–Milwaukee
Located in Room 168 of Lapham Hall
Department of Geosciences
Milwaukee, WI 53201
414-229-4561
www4.uwm.edu/letsci/geosciences/greene_museum/

MILWAUKEE PUBLIC MUSEUM
800 West Wells Street
Milwaukee, WI 53233
www.mpm.edu

THE GEM SHOP, INC. (rock shop)
W64 N723 Washington Avenue
Cedarburg, WI 53012
262-377-4666

BURNIE'S ROCK SHOP
901 East Johnson Street
Madison, WI 53703
608-251-2601
www.burniesrockshop.com

Wisconsin (continued)

GARY'S ROCK SHOP
317 S Main Street
Viroqua, WI 54665
608-637-7700
www.garysrockshop.com

JACK PINE ROCK SHOP
15822 East 2nd St
Hayward, WI 54843
715-934-2130

Iowa Museums and Rock Shops

UNIVERSITY OF IOWA MUSEUM OF NATURAL HISTORY
17 North Clinton Street
Iowa City, IA 52240
319-335-0480
www.uiowa.edu/~nathist/index.html

FOSSIL & PRAIRIE STATE PARK
1227 215th Street
Rockford, IA 50468
641-756-3490
fossilcenter.com

GEODE STATE PARK (museum in visitors' center)
3333 Racine Avenue
Danville, IA 52623
319-392-4601
geode@dnr.iowa.gov

HONEY CREEK GEMS (rock shop)
1228 Washington Street
Davenport, IA 52804
563-324-6032

Iowa (continued)

WESTSIDE AGATES (rock shop)
2347 230th Street
Ames, IA 50014
515-292-3803

Illinois Museums and Rock Shops

WESTERN ILLINOIS UNIVERSITY MUSEUM OF GEOLOGY
Located on the first floor of Tillman Hall
Department of Geology
115 Tillman Hall
Macomb, IL 61455
309-298-1151
www.wiu.edu/cas/geology/museum.php

THE FIELD MUSEUM
1400 S Lake Shore Drive
Chicago, IL 60605
312-922-9410
www.fieldmuseum.org

FRYXELL GEOLOGY MUSEUM
Augustana College
Located in Swenson Hall of Geosciences (near 38th Street)
Rock Island, IL 61201
309-794-7318

DAVE'S DOWN TO EARTH ROCK SHOP AND MUSEUM
704 Main Street
Evanston, IL 60202
847-866-7374

WOODIE'S ROCK SHOP
924 Broadway
Hamilton, IL 62341
217-847-3881

Illinois (continued)

DENNIS STEVENSON GEODES (pay-to-dig mine)
625 South 18th St
Hamilton, IL
309-337-3089

JACOBS GEODE SHOP & MINE (pay-to-dig mine)
823 East County Rd 1220
Hamilton, IL
217-847-3509

VICKERS GEODES (pay-to-dig mine)
810 Walnut St
Hamilton, IL
319-795-1219

Bibliography and Recommended Reading

Books about Wisconsin, Iowa and Illinois Minerals

Cross, Brad L. and Zeitner, June Culp. *Geodes: Nature's Treasures*. Baldwin Park: Gem Guides Book Company, 2006.

Dott, Robert H. Jr. and Attig, John W. *Roadside Geology of Wisconsin*. Missoula: Mountain Press Publishing Company, 2004.

Lynch, Dan R. and Lynch, Bob. *Agates of Lake Superior: Stunning Varieties and How They Are Formed*. Cambridge: Adventure Publications, Inc., 2011.

Robinson, George W. and LaBerge, Gene L. *Minerals of the Lake Superior Iron Ranges*. Houghton: Michigan Technological University, 2013.

General Reading

Bates, Robert L., editor. *Dictionary of Geological Terms, 3rd Edition*. New York: Anchor Books, 1984.

Bonewitz, Ronald Louis. *Smithsonian Rock and Gem.* New York: DK Publishing, 2005.

Chesteman, Charles W. *The Audubon Society Field Guide to North American Rocks and Minerals.* New York: Knopf, 1979.

Johnsen, Ole. *Minerals of the World.* New Jersey: Princeton University Press, 2004.

Mottana, Annibale, et al. *Simon and Schuster's Guide to Rocks and Minerals.* New York: Simon and Schuster, 1978.

Pellant, Chris. *Rocks and Minerals.* New York: Dorling Kindersley Publishing, 2002.

Pough, Frederick H. *Rocks and Minerals.* Boston: Houghton Mifflin, 1988.

Robinson, George W. *Minerals.* New York: Simon & Schuster, 1994.

Index

About the Authors

Dan R. Lynch has a degree in graphic design with emphasis on photography from the University of Minnesota Duluth. But before his love of art and writing came a passion for rocks and minerals, developed during his lifetime growing up in his parents' rock shop in Two Harbors, Minnesota. Combining the two aspects of his life seemed a natural choice and he enjoys researching, writing about, and taking photographs of rocks and minerals. Working with his father, Bob Lynch, a respected veteran of Lake Superior's agate-collecting community, Dan spearheads their series of rock and mineral field guides—definitive guidebooks that help amateurs "decode" the complexities of geology and mineralogy. He also takes special care to ensure that his photographs compliment the text and always represent each rock or mineral exactly as it appears in person. Encouraged by his wife, Julie, he works as a writer and photographer.

Bob Lynch is a lapidary and jeweler living and working in Two Harbors, Minnesota. He has been cutting and polishing rocks and minerals since 1973, when he desired more variation in gemstones for his work with jewelry. When he moved from Douglas, Arizona, to Two Harbors in 1982, his eyes were opened to Lake Superior's entirely new world of minerals. In 1992, Bob and his wife Nancy, whom he taught the art of jewelry making, acquired Agate City Rock Shop, a family business founded by Nancy's grandfather, Art Rafn, in 1962. Since the shop's revitalization, Bob has made a name for himself as a highly acclaimed agate polisher and as an expert resource for curious collectors seeking advice. Now the two jewelers enjoy a partial retirement, keeping Agate City open on the weekends while taking more time for family and traveling the rest of the year.

Notes